331

Nuclear Power and Public Policy

DATE DUE

JAN 3 0 1986			
NOV 0 2 1986			
MAR 1 6 1992			
NOV 1 5 1992			
OCT 2 0 1994			
DEC 3 0 1994			
DEC 2 9 RECD			
NOV 0 5 2004			
JUL 1 0 2005			
GAYLORD			PRINTED IN U.S.A.

A PALLAS PAPERBACK / 15

Nuclear Power and Public Policy

The Social and Ethical Problems of Fission Technology

by

K. S. Shrader-Frechette

University of California, Santa Barbara, California, U.S.A.

Second Edition, Revised

D. Reidel Publishing Company

Dordrecht : Holland / Boston : U.S.A.
London : England

Library of Congress Cataloging in Publication Data

Shrader-Frechette, K. S., 1944–
   Nuclear power and public policy.

   (A Pallas paperback ; 15)
   Includes bibliographical references and indexes.
     1. Atomic energy industries–United States. 2. Atomic energy–Social
aspects–United States. 3. Atomic energy–United States–Moral and ethical aspects.
4. Atomic power–Law and legislation–United States. I. Title.
HD9698.U52S54   1982   338.4′762131256′0973   82–20427
ISBN 90-277-1513-0 (pbk.)

Published by D. Reidel Publishing Company,
P.O. Box 17, 3300 AA Dordrecht, Holland.

Sold and distributed in the U.S.A. and Canada
by Kluwer Boston Inc., Lincoln Building,
160 Old Derby Street, Hingham, MA 02043, U.S.A.

In all other countries, sold and distributed
by Kluwer Academic Publishers Group,
P.O. Box 322, 3300 AH Dordrecht, Holland.

D. Reidel Publishing Company is a member of the Kluwer Group.

*First published in 1980 in hardbound edition (ISBN 90-277-1054-6) and in paperback
(ISBN 90-277-1080-5).*

Reprinted 1983 with revisions
Reprinted 1984 without corrections

For Maurice

Table of Contents

Preface

This book grew out of projects funded by the Kentucky Humanities Council in 1974 and 1975 and by the Environmental Protection Agency in 1976 and 1977. As a result of the generosity of these two agencies, I was able to study the logical, methodological, and ethical assumptions inherent in the decision to utilize nuclear fission for generating electricity. Since both grants gave me the opportunity to survey public policy-making, I discovered that there were critical lacunae in allegedly comprehensive analyses of various energy technologies. Ever since this discovery, one of my goals has been to fill one of these gaps by writing a well-documented study of some neglected social and ethical questions regarding nuclear power.

Although many assessments of atomic energy written by environmentalists are highly persuasive, they often also are overly emotive and question-begging. Sometimes they employ what seem to be correct ethical conclusions, but they do so largely in an intuitive, rather than a closely-reasoned, manner. On the other hand, books and reports written by nuclear proponents, often under government contract, almost always ignore the social and ethical aspects of energy decision-making; they focus instead only on a purely scientific assessment of fission generation of electricity. What the energy debate needs, I believe, are more studies which aim at ethical analysis and which avoid unsubstantiated assertions. I hope that these essays are steps in that direction.

xiii

Although I conclude that there are grave questions besetting nuclear fission, my arguments should not be taken to mean either that there are no troubling problems inherent in other means of generating electricity, or that all proponents of atomic power are prey to dishonest or profit-seeking motives. On the contrary, I have found both that all energy technologies have myriad economic and social costs, and that most of the supporters of nuclear fission are sincere, hardworking people who are interested in the well-being of this country and its people. If implementation of nuclear technology is ethically questionable today, then it is in part because its earliest proponents did not have the benefit of hindsight, and because they were overly optimistic regarding the success of their quest for cheap, clean, abundant energy, but not because they have all been evil or compromising persons. It is easy to criticize with hindsight. To the extent that there is something of value in these pages, it is because the questions raised here have become more clearly understood than they were two decades ago, when commercial generation of nuclear power began.

If the problems addressed in this book are resolved, then I will count myself among the proponents of nuclear fission. We ought not to align ourselves, *a priori*, against any technology and, as a culture, we cannot regress to pre-technological times. I have little sympathy for social critiques of a Luddite vein, although I believe we do need both to waste less energy and to live more simply. We also ought to engage in a more complete, democratic, ethical assessment of any technology before we decide to make it part of our public policy. Most importantly, we need a comprehensive analysis of all the various energy sources. If we do not accomplish such an assessment, we may be led to implement a particular technology, e.g., nuclear fission, even though its assets are in fact less desirable than those of its alternatives. On the other hand, we may be led to condemn a particular technology, e.g., solar, even though its liabilities are fewer than those of its competitors. With energy policy, as with life, there are no panaceas and no zero-cost alternatives. The least-cost options, however, can only be discovered after a thorough study of all possible courses of action. Since there has not been room to present such an assessment in this book, any final verdict on the desirability of employing nuclear fission is contingent both upon complete analyses of all other energy options and upon a comparison of those results with ones obtained in this and other studies of atomic power.

A shorter, earlier version of Chapter Two appeared in *The Journal of American Culture*. I am grateful to David Wright and Bob Snow who offered excellent suggestions for improving this essay. Chapter Four grew out of a brief presentation in 1978 at an energy symposium directed by Galen Renwick at Indiana University Southeast; an early, written version of this talk was published in *Research in Philosophy and Technology*. Approximately one-third of Chapter Five was presented in 1979 in an address at Michigan State University. Finally, Chapter Six grew out of a short paper which was presented in 1978 at Georgia Institute of Technology, revised, and given again in 1979 at the University of Pittsburgh at Johnstown. I am grateful for all these opportunities and for all the people who helped me to present, clarify, and amend my ideas.

I would like to thank especially my husband, Maurice, the brightest and most loving critic my work has; Ken Sayre, my graduate school mentor; my father, Owen Shrader, who taught me a reverence for nature; my mother, Mildred Shrader, whose goodness inspires me; and all the members of the Philosophy Department of the University of Louisville for their support, their perceptive intellects, their tolerance, and their good cheer. I am also grateful to Betty Myers Shrader, my typist, for her precision, thoroughness, and sense of humor; to my graduate and undergraduate students, from whom I have learned much; and to many environmentalist friends, like Fred Hauck, whose selfless dedication to social justice and ecological well-being has encouraged me. Finally, I thank Lawrence Badash for his helpful referee's comments and L. K. Richardson for her thoughtful and efficient editorial guidance.

The University of Louisville   K.S.S.-F.
August, 1979

Preface to the 1983 Edition

"The utility lied . . . the NRC covered up to protect the nuclear industry . . . the media engaged in an orgy of sensationalism." These are the fragments of the controversy recorded in one of the 1979 reports of the President's Commission on the Accident at Three Mile Island (Task Force Report, *The Public's Right to Information*, p. 3). As a result of the Pennsylvania incident, nuclear proponents have taken to claiming: "see, 'the system' works; we didn't have a full core melt". Equally vociferously, opponents argue: "the Three Mile Island (TMI) crisis showed that a major catastrophe is highly probable". Examining both positions, John Kemeny, Chair of the President's Commission to investigate the accident, revealed (before the Subcommittee on Energy Research and Production of the Committee on Science and Technology of the US House of Representatives (p. 7)): "an accident like TMI was inevitable".

Since the first edition of *Nuclear Power and Public Policy* was written several months prior to TMI, this new edition provides an opportunity to reflect on the policy consequences of the March 1979 crisis. If anything, this most serious accident in the history of US commercial nuclear power has refueled the debate over using fission to generate electricity. On the one hand, the March 1980 report of the Committee on Science and Technology of the US House of Representatives adopted the stance of "blaming the victim". It said that "the most significant after effects of the

xvii

[TMI] accident ... resulted from the ... general lack of understanding of the risks associated with nuclear power" (p. 7). After all, they said, TMI "was most definitely not a disaster" (p. 8). And, in March, 1982, in *US Commercial Nuclear Power*, the US Department of Energy predicted a steady increase in nuclear (fission) generating capacity to the year 2020, when the nuclear capacity would be 500 percent of the 1981 level (p. 37). On the other hand, Stuart Udall claimed (before the Committee on Interior and Insular Affairs of the House of Representatives, p. 2) in 1981 that the proverbial ant, a tiny malfunctioning valve at TMI, had brought the elephantine nuclear industry to its knees; the accident, he said, has "shaken the economic foundation of the entire electric utility industry" and generated more than $ 1 billion in cleanup costs and more than $ 24 million in replacement power costs.

Interestingly, the TMI accident reinforces many of the ethical reasonings I apply to the earlier history of commercial generation of electricity in the US. In Chapter 2, I criticize repeated assent, throughout the history of the nuclear debate, to the assumption that what is normal is moral; yet the assumption appears again, in discussions of the effects of TMI radiation exposures. The Subcommittee on Energy Research and Production of the Committee on Science and Technology of the US House of Representatives (p. 8) dismissed the exposures on the grounds that they were "substantially less than that received from a number of other sources". (That claim, of course, is tantamount to saying that any injury is negligible so long as it is less than that 'naturally' occurring from other causes.)

In Chapter 3, I show how government subsidy of waste disposal skews the methodology of cost-benefit analyses of nuclear fission; yet, in the aftermath of TMI, the Department of Energy subsidized the cleanup with $ 100 million of its funds. The result is that the costs of TMI and the consequences of those costs are not likely to be incorporated into industry assessments of whether nuclear fission is financially viable.

Likewise I claim in Chapter 4 that nuclear industry officials have often employed an argument from ignorance; they alleged, after repeated accidental releases of radiation, that there was "no health hazard to the public", even though there were no monitors working in many of the accident situations. The same happened at TMI; industrial and government officials repeatedly took the absence of concrete information as confirmation that exposure levels were low, as 1979 US hearings (by the Subcommittee on Natural Resources and Environment of the Committee on Science and Technology of the US House of Representatives, p. 3) reveal.

In Chapter 6, I argue against repeated claims that "statistically insignificant" deaths or "statistical casualities" are negligible. Yet, almost typically, in its 1980 report on TMI, the US General Accounting Office maintained that "no one died as a result of this accident" and that "the radioactive releases from the accident will have no or negligible effect on the physical health of individuals" (p. 11). However, even the most conservative government estimates (e.g., those of the US House of Representatives' Subcommittee on Energy Research and Production, in their 1980 report to the Committee on Science and Technology, p. 6) calculate that at least one or two statistical casualities can be expected from TMI.

The recurrence of these typical themes indicates, I think, that whether one speaks of incidents in 1979, prior to TMI, or in 1982, posterior to it, the *philosophical* problems persist and fuel the nuclear controversy. They are issues central to the ethical and methodological analysis of the underpinnings of all technology, whether solar, or coal, or fission, or some other. Hence, one ought not to take the easy path of assuming either that these ethical difficulties face only nuclear fission, or that there are any zero-cost options in any technology. A sound energy policy cannot be based on a few criticisms of the methodology of nuclear assessments. For that, something beyond this book is required. At the least, a comparative analysis of all energy technologies is needed. But before that analysis is possible, a clear criticism of the nuclear option must be available. I trust that *Nuclear Power and Public Policy* is one step in that direction.

The University of California   K.S.S.-F.
Santa Barbara
July 10, 1982

Introduction

The American people are ambivalent about the role government ought to play in society, says Bruce MacLaury, President of the Brookings Institution. On the one hand, they often approve increased provision of government protections and services, while on the other hand they disapprove "the proliferation of bureaucracy".[1] For this reason, questions of public policy are usually caught on the horns of a dilemma.

Perhaps nowhere is the public policy dilemma so fraught with difficulties as it is in the case of choosing acceptable energy technologies. Many Americans want the government to guarantee continuing supplies of electrical power and to insure an expanding economy often associated with high energy consumption. At the same time, they want both to be protected from the health hazards associated with most means of generating electricity and to be assured an inexpensive price for whatever power they do use. Clearly government cannot do all things. It cannot, for example, promote advanced energy technologies whose safe operation demands constant monitoring, and at the same time curtail the regulatory and rule-making functions of agencies of big government. Hence the people must make their choices. They must assess the acceptability of alternative means of generating electricity, and they must decide what political, moral, and environmental prices, if any, they are willing to pay to employ various energy technologies.

1

'Technology assessments', the new order of the day, ought to provide a basis for articulating viable public policy regarding energy strategies. In many cases, however, these analyses have focused on economic and scientific factors associated with a particular technology but, because of their difficulty, have ignored investigations of relevant social, political, and ethical parameters. Understandable as it is, this omission has been the source of much controversy regarding nuclear power. The failure of the US Atomic Energy Commission and the US Nuclear Regulatory Commission to conduct wide-ranging assessments of fission technology is well-known. In fact, one of the top federal commissioners charged with evaluating and regulating atomic power said recently that he wanted "to eliminate . . . from the public debate over nuclear energy", the "extraneous" and "irrelevant" ethical, social, and political issues "which cloud . . . meaningful dialogue".[2]

In large part, this elimination has already been accomplished. Within the federal evaluative and regulatory process, as will be seen later, there is not, and has never been, a forum for the public to express its views on the policy, as opposed to the scientific, questions raised by nuclear power. Moreover, beginning in 1957, approximately 60 reactors were built and licensed to generate electricity in the US before a complete evaluation of nuclear power safety was undertaken by the government. When the only allegedly complete study (WASH-1400, known as *The Rasmussen Report*) was released in 1975, it concluded that fission reactors presented only a minimal health threat to the public. Early in 1979, however, the Nuclear Regulatory Commission withdrew its support from WASH-1400. Despite the fact that there is no complete study of nuclear power now approved by the US government, this country currently has 135 reactors either in operation or under construction.[3] Although scientific and technological criteria are employed by the NRC in the licensing process, questions of nuclear policy have been, and remain, unaddressed by any comprehensive government analysis. The reluctance of federal officials to address complicated and controversial aspects of nuclear energy policy is understandable. However, even if a consensus regarding atomic power is not possible in this country, much is to be gained educationally and democratically by allowing the public to discuss all parameters relevant to the fission question.

The purpose of this book is to raise some of the social, political, and ethical issues which for so long have been ignored in making government assessments of nuclear power. In raising these questions, I hope to provide valuable insight and analysis necessary for articulating a viable public policy toward atomic energy. The aim is not only to establish ethical criteria for assessing nuclear technology but also to suggest some of the constraints that ought to operate in our numerous risk-benefit tradeoffs.

Chapter One, 'Nuclear Technology', first briefly outlines the history of using atomic fission to generate electricity and then describes the main features of a nuclear reactor and its principles of operation. The purpose of the historical and scientific material thus summarized is to provide a context for understanding the technology-related ethical and social issues raised later in the book.

Much of the debate over nuclear power has focused on the possibility of a catastrophic accident which (according to government estimates) could result in 45,000 immediate deaths.[4] Many of the arguments of nuclear proponents, however, are predicated on the assumption that such an accident is unlikely. Hence, in the minds of many persons, one of the most important assessments of nuclear technology ought to be based on the assumption that no major accidents will occur. Although I believe there are numerous reasons for questioning this assumption, it is employed for purposes of argument in Chapter Two. In this chapter, 'Reactor Emissions and Equal Protection', I presuppose that all reactors will operate normally. Given this assumption, I ask whether current public policy (governing admissible releases of radioactivity during generation of electricity) is based on sound ethical premises. Because of implicit violations of the right to equal protection, I argue that current policy regarding nuclear technology is unjust, even if one assumes that no accidents occur.

Besides emissions of low-level radiation, there are other considerations which indicate that, even under normal conditions, public policy regarding atomic fission generation of electricity poses grave ethical questions. One such consideration is radioactive waste. In Chapter Three, 'Nuclear Wastes and the Argument from Ignorance', I argue that it is ethically reprehensible to generate long-lived nuclear wastes without knowing whether these carcinogenic and mutagenic substances can be safely stored. Such a policy, I maintain, is based on the questionable presupposition that it is acceptable to contract a debt to future generations without knowing whether the debt can be paid.

In Chapter Four, 'Core Melt Catastrophe and Due Process', I present an ethical and methodological assessment of public policy based on the presupposition that a core melt accident is improbable. First, I argue that such an assumption is implausible on both logical and scientific grounds. Next, I show why current public policy regarding liability coverage in the case of a nuclear catastrophe violates constitutional principles of due process.

Despite the fact that my arguments regarding due process and equal protection are probably sufficient to reveal basic flaws in public policy regarding atomic fission, next I evaluate cost-benefit analyses of nuclear technology. In Chapter Five, 'Nuclear Economics and the Problem of Externalities', I argue that the alleged cost-effectiveness of fission generation of electricity is based on economic methodology which is both illogical and unethical. The methodology is deficient in that it ignores many 'externalities' or social costs of the nuclear fuel cycle.

In the final chapter, 'Nuclear Safety and the Naturalistic Fallacy', two well-known assessments of fission technology are evaluated.

Here, the focus is not on the policy problems generated by low-level radiation, limited nuclear liability, and the possibility of a core melt, but on the deficient ethical reasoning often underlying government assessments of nuclear power. In a general sense, this last chapter explains the faulty methodology responsible for the ethical assumptions criticized in previous chapters. It also outlines the sorts of policy-making procedures which ought to be followed in dealing with nuclear technology. Since the key to avoiding the ethical, social, and political problems outlined in the book is to revise our mode of making public policy, this last chapter closes with my own suggestions for reform. Without such changes, I argue, public policy will continue to prevent rather than to promote the realization of democratic ends.

Notes

[1] S.W. White, 'Public Policy and Private Interest', *National Forum* **LXIX**(1), (Winter 1979), 2.
[2] W.O. Doub, 'Meeting the Challenge to Nuclear Energy Head-On', *Atomic Energy Law Journal* **15**(6), (Winter 1974), 261, 263.
[3] This statistic was provided by the Council on Environmental Quality, *The Good News About Energy*, US Government Printing Office, Washington, D.C., 1979, p. 21.
[4] For discussion of these and other calculations contained in WASH-740, see Chapter Four of this volume.

Chapter One

Nuclear Technology

'Nuclear technology' is a broad term encompassing both fission and fusion. Since the only type of nuclear power currently used for commercial generation of electricity is fission, this analysis is limited to a discussion of fission technology.

Today, 19 nations are producing electricity by means of atomic energy, and at least 20 developing countries have embarked on nuclear power programs.[1] In the US alone, as of December 31, 1981, 151 reactors were operating or under construction.[2] Government experts predicted in 1973 that there would be 1000 fission plants generating power in the US by the year 2000, and that by that time 60% of the nation's electricity would be supplied by nuclear energy; by 1982, their predictions were for 250 plants by the year 2020, when nuclear fission would supply 32% of US electricity.[3] Even though reduced energy demand and increased opposition to atomic power have scaled down these estimates, nuclear fission remains an important part of current US energy policy. This is in part because only coal and atomic power are presently available sources for new, large-scale electric generating plants.[4] For all these reasons it is vitally important to understand the implications of current US public policy regarding nuclear technology. In order to evaluate the ethical, social, and political issues raised by this policy, however, one needs to have some perspective on nuclear technology.

The purpose of this chapter is to provide such a perspective. It includes a brief history of nuclear technology; an outline of US government regulation regarding it; a basic explanation of fission generation of electricity; and a summary of the ethical problems raised by employment of nuclear technology.

1. The History of Nuclear Energy

The history of nuclear fission technology, widely employed in medicine and in the generation of electricity, is in large part the history of the atomic bomb. Both nuclear weapons and reactors produce energy by fissioning uranium or plutonium. In a fission reactor, the energy release is completely controlled and used to produce electricity. In a weapon, the energy release is largely uncontrolled. Both uses of fission, however, produce substantial amounts of radioactivity.

Nuclear and radiation-related technology dates back to Roentgen's discovery of the X-ray in 1895 and Becquerel's discovery of natural radioactivity in 1896. The great step forward, however, came in 1905 with Albert Einstein's Special Theory of Relativity. Einstein proved that matter could be converted into energy. Although at the time there was little discussion of the theoretical possibilities, once accurate atomic weights became available in the 1920's and 1930's, we learned that certain atoms could be split or fused to convert part of their mass to energy. Largely because of Einstein's discovery, nuclear concepts were already well-developed by the late 1930's. When Fascism then drove Einstein, Fermi and Szilard to the US, they continued their research on nuclear fission. Afraid lest Hitler be the first to develop an atomic bomb, Einstein wrote President Roosevelt and told him it might be possible for the US to build such a weapon.

In 1942, the Manhattan Engineer District Project was formed; and, in December of the same year, a team directed by Enrico Fermi produced the world's first nuclear chain reaction. By the next month, the federal government was overseeing the building of the first atomic bombs at Oak Ridge, Tennessee, and Hanford, Washington. The Oak Ridge scientists had the task of producing bomb-grade U-235. The nuclear reactors at Hanford used natural uranium to produce the plutonium needed for nuclear warheads. On July 16, 1945 the world's first atomic bomb, using plutonium, was exploded in new Mexico. On August 6, 1945 the first nuclear warhead, employing U-235, was dropped on Hiroshima; some 65,000 people perished in its blast.

From 1940 through 1945 the US spent $ 2 billion to develop the first atomic bombs used during World War II. Thereafter the government took twenty years and more than $ 100 billion in subsidies to develop the first power reactors used to generate electricity. The reasons for beginning to develop fission reactors in the 1940's and 1950's were that the military wanted bombs and the government hoped to take advantage of its new technology for peaceful, as well as wartime, purposes. Scientists were optimistic about the "Atoms for Peace" program; it also provided a non-warlike rationale for continuing the development of nuclear energy. Prodded on, both by the escalating Cold War and by hope in the peaceful atom, government was able to develop commercial reactors and, at the same time, obtain the weapons-grade plutonium as a reactor by-product.[5] Carl Walske, who holds a Ph.D. in physics as well as the presidency of the coalition of pro-nuclear industries, the Atomic Industrial Forum, has pointed out repeatedly that military expenditures and justifications were ultimately responsible for the development of nuclear technology.[6]

In 1946, Congress enacted the nation's first Atomic Energy Act [42 U.S.C. § 2011], creating the Atomic Energy Commission and authorizing exclusive goverment control of peacetime uses of nuclear power.[7] By 1950, "the atomic bomb program had become ... the country's largest industrial enterprise". In 1953, for example, the US Atomic Energy Commission owned and operated three towns, employed 5% of the nation's construction labor force, and consumed 10% of the country's electric power.[8] This military development of atomic energy was considered necessary, since the USSR had already tested its own nuclear devices, and the US was engaged in the beginning of the Cold War.

One legacy of the arms race of the forties and fifties is that US commercial nuclear technology is built upon US military nuclear technology. As a consequence, the US employs a type of reactor much more susceptible to catastrophic accident. During the arms race following World War II, the US expended billions of dollars of research and development monies on water-cooled reactors because they were not complex to build and because their fuel was enriched uranium, already being used for making explosives. Enrichment plants existed for the bomb effort, and their continued operation could be justified if they also were used to make fuel for reactors. As a result, the US nuclear technology is built around a water-cooled, enriched-uranium design. This is a critical point, because such reactors (unlike those of Canada, Britain, and the USSR) are plagued with a much higher risk of 'core meltdown', the chief potential cause of catastrophic reactor accidents.

The key benefit of plants which do not use enriched uranium is that their fuel, natural uranium, contains only 0.7% uranium-235, and hence the density of power generation in their fuel is much less than that in enriched uranium. As a result, the natural uranium reactors can withstand a loss of cooling more easily than can the enriched uranium reactors. Cooling helps to remove the heat generated by fissioning uranium or plutonium which releases energy; without cooling, the uranium core could melt and cause catastrophic damage. Such an incident could release radiation equivalent to that of 1000 Hiroshima bombs.[9] According to US government estimates, were a core melt to occur, property damages alone could reach $ 17 billion, and an area the size of Pennsylvania could be destroyed.[10]

Such catastrophic accidents, however, have never occurred.[11]

This success is part of the reason why the transition from military to commercial use of nuclear technology has been so rapid. After many years of military development, the commercial phase of nuclear power was ushered in with the passage of the Atomic Energy Act in 1954. The purpose of this legislation was to promote the general welfare by development of nuclear technology "within the bounds of the national defense and the health and safety of the public".[12] Private utilities were given government subsidies to develop nuclear power plants, and in 1956 the AEC guaranteed that it would buy the plutonium these plants produced, since it was needed for the governmental development of nuclear warheads. By 1957 the nation's first commercial atomic plant (in Shippingport, Pennsylvania) was ready to begin operation.

Although national defense and the possibility of developing a cheap, clean, abundant energy source provided the ultimate rationale for beginning commercial development of nuclear power, the technology continued to expand, even after the US had more than enough weapons-grade plutonium as a reactor by-product. As a consequence, the US has sponsored a growing technology without ever assessing fully whether it is a desirable means to a somewhat different end, viz., electrical energy rather than nuclear warheads and "atoms for peace".[13]

2. Government Regulation of Atomic Power

In one sense, however, industry did evaluate the desirability of employing atomic energy to generate electricity. Its assessment has been the motivating factor in much US government regulation of nuclear fission. Power companies refused initially to invest their funds in the technology; they claimed that the high risks of atomic energy made it unlikely that nuclear generation of electricity could ever be profitable. Given the possibility of a core melt, owing to the use of enriched uranium, business and industrial leaders said unequivocally that they would not invest in nuclear power because "the legal consequences of a serious accident would financially destroy a company".[14] Every major corporation with nuclear interests said, through their spokespersons (e.g., the Vice-President of General Electric), that "the company would withdraw from the nuclear field" unless some indemnity legislation was passed to protect the industry.[15] Thus, in addition to the passage of the 1954 Atomic Energy Act, whose object was "to promote the private development of atomic energy",[16] three years later Congress passed the Price-Anderson Act. An amendment (Section 170) to the earlier act, the Price-Anderson legislation contains a number of provisions. The most important of these guarantees that the law will "hold harmless the [nuclear] licensee and other persons indemnified" from public liability claims arising from nuclear accidents causing total damages in excess of $ 560 million.[17]

The liability limit provided by the Price-Anderson Act was passed in 1957 with the proviso that the limitation would be effective for only ten years. The Joint Committee on Atomic Energy assured Congress that the limitation was needed only for a decade, since, by 1967, they said, "the problems of reactor safety will be to a great extent solved".[18] They argued that, by this date, the safety of nuclear plants would be evident and would "remove the fear of public liability that had previously blocked private investment".[19]

Despite the fact that the Price-Anderson Act was intended merely as a temporary measure, Congress was forced to extend it in 1965 for another ten years.[20] In 1975, the temporary act was again extended for a ten-year period.[21] Government officials, environmental writers, and attorneys for nuclear utilities have all made it clear that the US was forced to extend the Price-Anderson Act as a necessary condition for industry's generation of nuclear power.[22]

In 1975, the year in which the Price-Anderson Act was granted

its second extension, federal regulatory control of nuclear technology was also modified through the Energy Reorganization Act.
After thirty years during which the same government agency, the
Atomic Energy Commission, had both promoted atomic industry
and regulated it, the nuclear promotional and regulatory functions
of government were separated. The Energy Reorganization Act
abolished the AEC. At the same time, it created the Energy Research and Development Agency (to promote nuclear energy) and
the Nuclear Regulatory Commission (to assess the safety and
security of existing and proposed nuclear installations). The reorganization act became necessary because of numerous lawsuits
against the Atomic Energy Commission. The AEC was repeatedly
censured by the courts and the public for failing to regulate safety
hazards of nuclear power plants, and for subscribing to industry's
demands rather than the public's best interests.[23] The successor to
the AEC, the NRC (Nuclear Regulatory Commission) was charged
solely with regulating and licensing nuclear plants; the new body
consists of five commissioners, each appointed by the President
with the advice and the consent of the Senate. NRC regulation of
nuclear technology does not only include licensing fission reactors
and insuring their safe operation. It also involves setting standards
for allowable releases of radiation throughout all the processes
necessary to generation of electricity by means of atomic power.

3. Fission Generation of Electricity

To understand the significance of government regulation of nuclear
technology, it is necessary to know something about the operation
of fission reactors. This, in turn, requires a few facts about the
fission process, its supporting fuel cycle, its normal radioactive
discharges, and its potential for catastrophic accidents.

In the process known as nuclear fission, enormous amounts of
heat (energy) are released when uranium-235 atoms are bombarded

by neutrons, absorb neutrons, and split into lighter elements like strontium and iodine. When the uranium splits, other neutrons are also released which repeat the process in a chain reaction. As a result of this fissioning, many unstable elements are created, i.e., they lose energy by emitting particles. These energy losses are measured as radioactive emissions; in large enough amounts, radiation can cause death, cancer, genetic damage, or numerous other injuries.

Despite the dangers posed by radiation, nuclear technology has risen to great importance because of the immense energy that can be released from a small volume of highly processed uranium-235. Fissioning one ounce of U-235 produces as much heat as burning 100 tons of coal.[24] This means that if enough high-grade uranium ore could be found, one might be able to employ nuclear fission technology and thereby avoid all the water pollution, black-lung disease, and mining accidents associated with using coal to generate electricity. Also, since nuclear plants do not burn hydrocarbons, as oil and coal plants do, they do not produce conventional forms of air pollution. Instead, during normal operation, reactors emit small amounts of radioactive materials into the air and water near the plant.

All US reactors used commercially to generate electricity are known generically as light-water reactors (LWRs), since they use light water to slow down (or 'moderate') the fission process.[25] The heat released during fissioning is used to turn water into steam, and when the steam is directed against the blades of a turbine, electricity is created by rotating a generator. In other words, once heat is generated in the nuclear reactor, it is used to create electricity in the same manner as in a coal, gas, or oil plant.

Approximately 100 tons of cylindrical uranium fuel pellets are inside the fission reactor. These pellets are packed into rods and loaded in such a way that a controlled amount of fissioning can occur and water can circulate among the rods. The fuel load is surrounded by water (under pressure) which cools the fuel and slows down the fission process. The two basic types of LWRs differ, in part, in the degree of pressure under which this water is held. Boiling water reactors (BWRs) contain water under 1000 pounds per square inch, while pressurized water reactors (PWRs) contain water under 2250 pounds per square inch.[26]

In order to control the fission process in either a BWR or PWR, long rods of neutron-absorbing boron or cadmium are placed in the reactor among the bundles of fuel rods. Called 'control rods', they can be raised or lowered to regulate the rate of the fission process. Even after control rods have been inserted to stop the nuclear chain reaction, however, the residual radioactivity in the core can generate enough heat to melt the core. This is why operation of a cooling system is always necessary in a LWR.

If something prevents normal operation of the cooling system, the LWR is designed so that a backup system will flood the core. If this second system does not operate, then the emergency core cooling system (ECCS) is required to reflood the reactor with more water. If the electricity fails, however, the ECCS cannot operate, and the core melts. Government studies have indicated that such a melt "would likely lead to a failure of the containment".[27] This, in turn, would mean that an area of 40,000 square miles could receive radioactive contamination.[28] The worst type accident, say government experts, would result from a steam explosion in the reactor vessel; this would cause the containment vessel to rupture and release radioactivity into the surrounding air. Because of the magnitude of radioactivity that could be released, they say, every member of the entire population of the US is close enough to some reactor site "to receive some of the . . . dose calculated".[29] In the event that such a catastrophe occurred, says the government, approximately 45,000 people could be killed instantly and 100,000 could die as a result of the accident.[30]

Since the radioactive inventory of a reactor is larger than that of a bomb (i.e., it is equal to that of approximately 1000 Hiroshima-type weapons),[31] a reactor meltdown could be far worse than a bomb blast. (On the other hand, of course, the bomb would also produce deaths from the blast and from the high levels of initial radiation.) Because of the large amounts of radiation capable of being released from a reactor, much of the controversy over nuclear power has focused on the adequacy of the ECCS.

In addition to questions regarding the ECCS, however, a number of issues have been raised about other aspects of the 'nuclear fuel cycle'. To understand why this is so, let us look at the nuclear fuel cycle.

The term 'fuel cycle' refers to all the steps necessary both to prepare uranium for loading in a reactor and to store it after it has been used. These processes include mining, milling, conversion, enrichment, preparation and fabrication, reprocessing spent fuel, waste storage, and transportation among fuel cycle steps.

The mining stage of the uranium cycle is similar to that of the coal fuel cycle; in fact, government scientists estimate that workers in the uranium mining industry sustain 'the same risk' of fatal and nonfatal injury as workers in the coal mining industry.[32] The causes of death and injury in each case are quite different, however. The uranium miner, for example, is exposed to the decay of radon and faces the possibility of carcinogenic or mutagenic injury. The coal miner, on the other hand, is exposed to coal dust and faces the possibility of developing black lung.

Since uranium ore contains 99.8% waste material, and because only about four pounds of the necessary U_3O_8 are present in one ton of ore, milling is necessary. This step of the fuel cycle involves ore preparation, concentration, and product recovery. Its end result is 'yellowcake', which is about 80% U_3O_8. In the milling process, radon gas is discharged through the mill effluent stack and small amounts of uranium, thorium, and radium are carried offsite in liquid effluents. Large quantities of 'tailings', insoluble materials left after the U_3O_8 has been extracted, are also produced; these

must be kept secure from the biosphere since they contain dangerous radionuclides 'in practically the same concentration' as in the
original ore.[33]

The third stage of the fuel cycle, conversion, involves producing
gaseous uranium hexafluoride (UF_6) from U_3O_8. (This step also
results in discharging small amounts of uranium to waterways.)
The UF_6 is then shipped to gaseous diffusion plants where it is
used to enrich uranium so that it is 2 to 4% (by weight) U-235.
As a consequence of the enrichment process, small amounts of
uranium are released to the atmosphere both through gaseous
emissions and through liquid effluents from a holding pond.[34]

In the preparation and fabrication stages, steps four and five of
the nuclear fuel cycle, the enriched UF_6 is converted to uranium
dioxide pellets which are placed in the zirconium alloy tubes com-
prising the finished fuel assembly. As in the other steps of the fuel
cycle, small amounts of uranium are released to the atmosphere
through gaseous emissions and liquid effluents, and a significant
amount of radiation-contaminated waste is generated.[35]

Once the fuel is fabricated, it is ready to be used in electrical
generation. It is during the fissioning process that defects in the
fuel cladding begin to develop, allowing some fission products to
escape to the coolant water and ultimately to the environment.
The fissioning process also creates some very hazardous radioactive
wastes. The magnitude of these wastes can best be understood by
noting that a LWR must annually discharge approximately one-
third of its fuel in order to make room for the necessary fresh
fuel.[36] The highly toxic spent fuel is then ready for the next stage
of the nuclear cycle, reprocessing.

When the fuel is removed from the reactor core, it is placed in
on-site storage pools, after which it is supposed to be shipped to
reprocessing plants. The purpose of reprocessing is to recover the
uranium and plutonium from the waste materials so that the former products can be recycled and used again as fuel.

Although the reprocessing step is allegedly the seventh stage of the nuclear fuel cycle,[37] there is no commercial fuel reprocessing occurring in the US at present. One reprocessing plant (Nuclear Fuel Services in West Valley, New York) operated for a time, but it was closed in 1971 because it emitted excessive amounts of radioactive gases and liquids, and contaminated air, water, fish, and wildlife for many miles surrounding the plant.[38]

Faced with skyrocketing costs and an inability to contain dangerous radiation emitted during reprocessing, no companies are willing to continue the attempt to reprocess stored fuel. General Electric has abandoned its partially-completed reprocessing plant in Illinois; Allied-General has asked the government to take over its unfinished South Carolina reprocessing plant; and Getty Oil has been trying unsuccessfully for almost a decade to sell its closed plant in West Valley.[39]

As a consequence of the failure to achieve reprocessing, two additional problems have beset the nuclear fuel cycle. First, because the hoped-for reprocessing technology has been discovered to be impractical and perhaps impossible, numerous reactors have nowhere to store their highly dangerous spent fuel. The Nuclear Regulatory Commission originally said that because of the toxicity of the spent fuel, it should be shipped away for reprocessing "within six to nine months after removal from the reactor".[40] Since there is nowhere to send it, however, 12 nuclear reactors currently "do not have the spent fuel storage capacity for a full core discharge".[41] By the end of 1979, 27 reactors will not have discharge space, and by the end of 1980, 37 will have no space.[42] The lack of space for a core discharge is a violation of NRC policies since, if an accident were to occur, there might be a need to unload the core and place it in storage pools so as to prevent a catastrophe.[43]

The second problem with US inability to achieve reprocessing, and the consequent failure of many reactors to have core discharge space, is that enormous quantities of dangerous radioactive wastes

(spent fuels) are being improperly stored. They are being trans-
ferred to temporary basins near reactors still having space,[44] instead of being sent to permanent storage facilities.

Permanent storage of dangerous radioactive wastes is the last
stage of the nuclear fuel cycle. After removal from the reactor (or,
as planned earlier, after reprocessing), spent fuel wastes are supposed to be sent to a permanent repository. This storage is necessary since the decay times of many of the radioactive elements
produced by the fission process are 'millions of years'; until this
decay is complete, radioactive wastes have the potential for causing
death, cancer, and genetic damage.[45] In fact, the current US inventory of radioactive waste would be enough to kill several billion
people, if it were to be dispersed throughout the environment.[46]

Currently, the US has no permanent waste storage facility for
spent reactor fuels. The technological problem of isolating such
wastes in perpetuity, so as to safeguard man and the environment,
is a difficult one. The government predicted in 1978 that, by 1995,
it would have the first experimental model of such a storage facility ready for testing.[47] In the meantime, great quantities of radioactive waste are being stored temporarily at various sites around
the country. Most of these highly dangerous, commercial radioactive wastes are stored in Hanford, Washington. To date, over
500,000 gallons of highly toxic wastes have leaked from storage
tanks at Hanford and have added measurable radiation both to the
Columbia River and to the Pacific Ocean.[48] In order to store these
wastes more safely, scientists have proposed placement in mined
repositories, deep ocean sediments or drill holes, and ejection into
space.[49] So far, however, no technology has proved satisfactory
for this purpose.

4. Ethical Problems Raised by Nuclear Technology

As this brief discussion of the fission process and the nuclear fuel cycle indicates, there are numerous ways in which humans might be exposed to dangerous levels of radioactivity, whether from normal emissions and effluents of the fuel cycle or from the possibility of an extraordinary nuclear catastrophe. All these avenues of hazard mean that we as a people must make some well-thought-out choices regarding the risk-benefit tradeoff of nuclear technology. Our choices are all the more important because, as I will argue throughout the ensuing chapters, government assessments of atomic energy have characteristically dealt only with scientific and economic parameters, despite the fact that conscientious researchers attempted to provide accurate, objective analyses. In ignoring important ethical, social and political considerations, many scientists have thereby omitted parameters bearing directly on democratic and equitable public policy.

Although this volume cannot deal adequately with all the topics that ought to be considered in assessing nuclear technology, it will raise a number of significant ethical questions regarding low-level radiation (Chapter Two), nuclear waste (Chapter Three), the possibility of a core melt (Chapter Four), and nuclear liability insurance (Chapter Four). In addition to these substantive points, the book will also address several methodological issues. The point of the latter investigations is to assess the adequacy of the methodology according to which public policy regarding nuclear technology is determined. Chapter Five will evaluate cost-benefit assumptions and Chapter Six will criticize methods of risk-assessment used to support current policy governing nuclear fission.

In the process of analyzing the six issues just mentioned, there are a number of other ethical questions on which I hope to shed some light. These include:

(1) How ought the government to restrict civil liberties in order to guard against sabotage, terrorist attack, illicit fabrication of weapons, or radioactive contamination during the lengthy nuclear fuel cycle?

(2) What health costs ought the public to pay for its energy demands?

(3) Will the greater good of future generations be served by continuing to deplete the world's finite supply of fossil fuels or by leaving a legacy of nuclear technology, with attendant radioactive wastes, genetic damage, and increased cancer rates?

(4) Does the consumer have the right to receive the cheapest form of electricity available, and if so, is this generated by nuclear fission?

(5) Will nuclear energy render the US energy-independent and therefore better able to exercise moral leadership in the world community?

(6) Is it morally and economically desirable to employ nuclear technology in order to encourage increasing levels of US energy consumption?

(7) Is a no-growth energy scenario ethically and economically desirable for underdeveloped nations?

(8) Ought the nuclear industry to equate 'energy demand' with 'energy need'?

(9) Is it ethically responsible to employ a technology before a means of storing its lethal wastes has been devised?

(10) Are financial and economic criteria sufficient to justify current atmospheric releases of carcinogenic, low-level radiation from nuclear reactors?

By coming to think more clearly about these sorts of questions, we increase the probability that public policy will be built on firm ethical foundations. Let us examine some of those foundations.

Notes

[1] M.A. Rowden, 'Nuclear Power Regulation in the United States: A Current Domestic and International Perspective', *Atomic Energy Law Journal* 17(2), (Summer 1975), 117.
[2] US Department of Energy, *US Commercial Nuclear Power*, US Govern-

ment Printing Office, Washington, DC, 1982, p. 17; hereafter cited as: DOE, *Commercial*. See also Council on Environmental Quality, *The Good News About Energy*, US Government Printing Office, Washington, D.C., 1979, p. 21; hereafter cited as: *Energy*.

3 P.L. Joskow, 'Commercial Impossibility, The Uranium Market, and the Westinghouse Case', *Journal of Legal Studies* 6(1), (January 1977), 165. Information on the 1000-plant prediction may be found in H.R. Price, 'The Current Approach to Licensing Nuclear Power Plants', *Atomic Energy Law Journal* 15(6), (Winter 1974), 230; hereafter cited as: 'Plants'. L.M. Muntzing, 'Standardization in Nuclear Power', *Atomic Energy Law Journal* 15(3), (Spring 1973), 22. The later prediction is found in DOE, *Commercial*, p. 37.

4 Saunders Miller, *The Economics of Nuclear and Coal Power*, Praeger, New York, 1976, p.v; herafter cited as: *Economics*. See also *Energy*, p. 21.

5 Sheldon Novick, *The Electric War*, Sierra, San Francisco, 1976, p. 22. This fact is also corroborated by M. Willrich, *Global Politics of Nuclear Energy*, Praeger, New York, 1971, pp. 5–6. Breakdown and documentation of the more than $100 billion cost of government development of nuclear fission technology is given by J.J. Berger, *Nuclear Power*, Dell, New York, 1977, pp. 94–147, esp. pp. 112, 144–47.

6 Cited by Novick, *Electric War*, pp. 32–33, in a taped interview with Walske.

7 L. Breitkopf, 'The National Environmental Policy Act of 1969 and Nuclear Plant Licensing', *Depaul Law Review* 26(3), (Spring 1977), 666; hereafter cited as: 'Nuclear Plant'.

8 Novick, *Electric War*, p. 38. See also US Atomic Energy Commission, *Comparative Risk-Cost-Benefit Study of Alternative Sources of Electrical Energy*, (WASH-1224), Government Printing Office, Washington, D.C., December 1974, p. 3-49; hereafter cited as: 'WASH-1224'.

9 Novick, *Electric War*, pp. 160–61; and Berger, *Nuclear Power*, p. 36.

10 For more explanation of core melt and its consequences, see Chapter Four of this volume.

11 Numerous deaths and injuries have resulted, however, from both commercial and governmental use of nuclear fission; for an account of these accidents, see Chapter Four of this volume.

12 Breitkopf, 'Nuclear Plant', 666.

13 E.D. Muchnicki, 'The Proper Role of the Public in Nuclear Power Plant Licensing Decisions', *Atomic Energy Law Journal* 15(1), (Spring 1973), 39, makes this same point. See also Novick, *Electric War*, p. 22.

14 J. Marrone, 'The Price-Anderson Act: The Insurance Industry's View', *Forum* XII(2), (Winter 1977), 607; hereafter cited as: 'Insurance'.

15 W.S. Caldwell *et al.*, 'The "Extraordinary Nuclear Occurrence" Threshold and Uncompensated Injury Under the Price-Anderson Act,' *Rutgers-Camden Law Journal* VI(2), (Fall 1974), 379; hereafter cited as: 'Nuclear Occurrence'.

[16] J.R. Brydon, 'Slaying the Nuclear Giants', *Pacific Law Journal* **VIII**(2), (July 1977), 781; hereafter cited as: 'Nuclear Giants'.

[17] 'AEC Staff Study of the Price-Anderson Act, Part I', *Atomic Energy Law Journal* **16**(3), (Fall 1974), 220.

[18] R. Lowenstein, 'The Price-Anderson Act', *Forum* **12**(2), (Winter 1977), 597; hereafter cited as: 'P-A'.

[19] Caldwell, 'Nuclear Occurrence', 364.

[20] In 1965 The Price-Anderson Act was amended so that a waiver of defenses against all liable persons was required, provided that an "extraordinary nuclear occurrence" had taken place. The Atomic Energy Commission was given the power to define occasions on which such an occurrence had happened, and defenses were waived for these incidents; the 'injured' person was required only to show that an accident at a nuclear power plant caused his injury, but not that the owners/operators were negligent. Amendments at this time also provided for immediate partial compensation to the victims without their having signed a release. (Marrone, 'Insurance', 608–9.)

[21] At this time Congress also passed two amendments to the act. It provided a deferred premium plan according to which, after approximately twenty-five more plants have been built, industry-financed nuclear insurance will be substituted for government coverage. A second amendment stipulated that, if approximately twenty-five more atomic plants are built, and if the maximum ($ 5 million per licensee) premium is assessed, then the nuclear industry's liability might rise (at most) at a rate of $ 5 million for each plant built after the next twenty-five. (More details regarding this plan will be given in Chapter Four; see also Lowenstein, 'P-A', 599–603; Marrone, 'Insurance', 608–609.)

[22] J.W. Gofman and A.R. Tamplin, *Poisoned Power*, Rodale, Emmaus, Pennsylvania, 1971, pp. 32, 180, 239 (hereafter cited as: *Power*), make this point, as do attorneys for nuclear utilities, e.g., R. Lowenstein, 'P-A', 596. See also M. Gravel, 'Price-Anderson: A No-Fault Nightmare,' in *Countdown to a Nuclear Moratorium* (ed. by R. Munson), Environmental Action Foundation, Washington, D.C., 1976, p. 48 (hereafter cited as: *Countdown*); A.W. Murphy and D.B. LaPierre, 'Nuclear "Moratorium" Legislation in the States and the Supremacy Clause', *Environment Law Review 1977* (ed. by H.F. Sherrod), Clark Boardman, New York, 1977, p. 405 (hereafter cited as: 'Nuclear Legislation'); and M.C. Olson, *Unacceptable Risk*, Bantam, New York, 1976, p. 55. Even government proponents of nuclear energy clearly reveal that industry would not 'go nuclear' without liability protection; see 'AEC Staff Study of the Price-Anderson Act, Part I', p. 209 and 'AEC Staff Study of the Price-Anderson Act, Part II', *Atomic Energy Law Journal* **16** (4), (Winter 1975), 298.

[23] The Energy Reorganization Act, as a response to government problems

regarding the AEC's pro-industry bias, is discussed in M.A. Rowden, 'Nuclear Power Regulation in the United States: A Current Domestic and International Perspective', *Atomic Energy Law Journal* 17(2), (Summer 1975), 102ff. See also Novick, *Electric War*, pp. 354–55; Brydon, Nuclear Giants, p. 771; V. McKim, 'Social and Environmental Values in Power Plant Licensing', in *Values in the Electric Power Industry* (ed. by K.M. Sayre), University Press, Notre Dame, 1977, pp. 38–40 (hereafter cited as: *Values*); G.B. Karpinski, 'Federal Preemption of State Laws Controlling Nuclear Power,' *Georgetown Law Journal* LXIV(6), (July 1976), 1336; and J.G. Palfrey, 'Energy and the Environment: The Special Case of Nuclear Power', *Columbia Law Review* LXXIV(8), (December 1974), 1380.

24 J.J. Berger, *Nuclear Power*, p. 32.

25 'Light water' is so named because of its relatively low concentration of deuterium, the hydrogen isotope having a neutron as well as a proton in its nucleus; 'heavy water' has a higher concentration of deuterium.

26 Berger, *Nuclear Power*, pp. 32–34; and AEC, 'WASH-1224', pp. 3-50 through 3-51, explain the fission process and the differences between BWRs and PWRs.

27 AEC, *Reactor Safety Study: An Assessment of Accident Risks in US Commercial Nuclear Power Plants* (WASH-1400), Government Printing Office, Washington, D.C., 1975, p.74.

28 AEC, 'WASH-1400', p. 172.

29 AEC, 'WASH-1400', pp. 117, 164.

30 These accident consequences were calculated in the government document WASH-740 and its update, i.e., US AEC, *Theoretical Possibilities and Consequences of Major Accidents in Large Nuclear Power Plants* (WASH-740), Government Printing Office, Washington, D.C., 1957; and R.J. Mulvihill, D.R. Arnold, C.E. Bloomquist, and B. Epstein, *Analysis of United States Power Reactor Accident Probability* (PRC R-695), Planning Research Corporation, Los Angeles, 1965; cited hereafter, respectively, as: 'WASH-740–57' and 'WASH-740'.

31 See note 9 above.

32 AEC, 'WASH-1224', p. 3-57.

33 AEC, 'WASH-1224', p. 3-60.

34 AEC, 'WASH-1224', pp. 3-60 and 3-61.

35 AEC, 'WASH-1224', p. 3-61.

36 Committee on Interstate and Foreign Commerce, *Nuclear Waste Management and Disposal*, Hearings Before the Subcommittee on Oversight and Investigations of the Committee on Interstate and Foreign Commerce, House of Representatives, 95th Congress, First Session, Serial 95-67, Government Printing Office, Washington, D.C., 1977, p. 1; hereafter cited as: Committee on IFC, *Nuclear Waste*.

[37] See AEC, 'WASH-1224', pp. 3-64 and 3-65 for a discussion of fuel reprocessing.

[38] See Berger, *Nuclear Power*, pp. 78—84.

[39] Berger, *Nuclear Power*, pp. 94—97.

[40] L.V. Gossick, Executive Director of the NRC, quoted in Committee on IFC, *Nuclear Waste*, p. 111.

[41] L.V. Gossick, NRC, quoted in Committee on IFC, *Nuclear Waste*, p. 115.

[42] ERDA-25 survey, cited in Committee on IFC, *Nuclear Waste*, p. 4.

[43] L.V. Gossick, NRC, quoted in Committee on IFC, *Nuclear Waste*, p. 101.

[44] ERDA-25 survey, cited in Committee on IFC, *Nuclear Waste*, pp. 4—5.

[45] J.M. Deutch and the Interagency Review Group on Nuclear Waste Management, *Report to the President by the Interagency Review Group on Nuclear Waste Management* (TID-2817), National Technical Information Service, Springfield, Virginia, October, 1978, p. iii; hereafter cited as: IRG, *Report*.

[46] US Energy Research and Development Administration, *Final Environmental Statement: Waste Management Operations, Hanford Reservation, Richland, Washington*, Vol. 1. (ERDA-1538), National Technical Information Service, Springfield, Virginia, October 1975, p. X-159; hereafter cited as: 'ERDA-1538', p. X-203.

[47] IRG, *Report*, pp. iv, 34. US ERDA, 'ERDA-1538', p. X-203.

[48] US ERDA, 'ERDA-1538', p. X-28.

[49] IRG, *Report*, pp. 25ff.

Chapter Two

Reactor Emissions and Equal Protection

A recent Army filmstrip, designed to explain military activities to civilians, described US employment of nuclear power as a "war against fear — fear of the unknown". The only way to dispel anxiety regarding this new technology unleashed at Hiroshima and Nagasaki, reasoned the author of the presentation, was to give the public "invaluable first-hand knowledge of the effects of atomic energy".[1]

In one way or another, thousands of Americans have obtained this "first-hand knowledge", either by witnessing weapons tests in Nevada, by living near a reactor, by transporting uranium and radioactive wastes, or by working in an atomic industry. Their experiences, however, have not dispelled fear of nuclear technology; if anything, they have fueled the accelerating debate over the peaceful employment of fission reactors to generate electricity. At the heart of this conflict are many important public policy issues, one of which is the subject of this chapter: Should the government continue to allow emission of low-level radiation during normal operation of the fuel cycle of nuclear power plants?

1. The Controversy over Low-Level Radiation

One reason why this particular public policy issue has become increasingly controversial is that the results of various statistical studies (of the effects of low-level radiation) are often in disagreement with each other. Scientific data used by one interest group is

25

frequently in direct conflict with that used by an opposed group. For example, the pro-nuclear government commission charged with regulating atomic plants calculated that 0.0002 cases of cancer per year would be induced by exposure to one rad of radiation.[2] Biophysicist Arthur Tamplin, nuclear chemist John Gofman and others, however, calculated that 32,000 cases of cancer per year would be induced by exposure to one rad.[3]

Perhaps the more important reason why public policy regarding low-level radiation has become increasingly debatable is that there is no real consensus on answers to basic ethical questions underlying the policy. There are a number of reasons why the lack of an ethical consensus to undergird current radiation policy contributes to public unease in this regard. For one thing, even when experts agree on the magnitude of a radiation risk, they disagree ethically on whether the risk ought to taken. In general the government takes the position that requiring 'zero risk' from radiation is not an acceptable ethical and regulatory position, since 'zero risk' would result in economically undesirable consequences.[4] Many environmentalists, on the other hand, argue that failure to require 'zero risk' from radiation is morally abhorrent because it "represents nothing other than a hunting license for human beings".[5]

Ethical disagreements in the low-level radiation controversy have surfaced, in particular, as a result of the National Research Council's 1972 study, *The Effects on Populations of Exposure to Low Levels of Ionizing Radiation*. The study simply assessed the risk, and then left it to the regulatory bodies to endorse exposure standards and therefore moral judgments regarding the acceptability of risk.[6] Likewise during the controversy two decades ago over the radiation risk from nuclear bomb tests, the key conflicts arose over values, not over the magnitude of risk. As Eugene Rabinowitch, biophysicist and former editor of *The Bulletin of the Atomic Scientists*, put it: "the adversary experts did not disagree on the facts [regarding radiation risk] of the situation; they disagreed only on moral conclusions [regarding the acceptability of the risk] which they derived from these facts".[7] Hence, apart from scientific disagreement over the magnitude of radiation risk, there is considerable controversy over the moral acceptability of a risk, even when its magnitude is defined or agreed-upon. This is why, apart from factual disagreements, radiation exposure standards "differ widely from country to country".[8]

Even within the US, the lack of ethical consensus regarding an acceptable risk complicates the debate over radiation standards. As one policymaker from the Environmental Protection Agency noted:

An ethical basis for decisions regarding risk transference is needed not only for philosophical reasons, but also for the practical purpose of implementing evaluation techniques such as cost-effectiveness and risk-cost analyses. Unfortunately, society has not established clear approaches for dealing with the imposition of such risks.[9]

According to what moral principles, for example, ought society to determine when the carcinogenic and mutagenic costs of low-level radiation have outweighed its technological benefits? What numbers of cancers and genetic deaths ought society to trade for the economic advantages of not requiring tighter radiation controls?

One reason why the low-level radiation controversy is of overarching importance is that it presents a paradigmatic case of a particular type of debate in the area of public policy governing technology. At the heart of this conflict are two problems: (1) how to employ risk/benefit analysis as a prelude to government regulation; and (2) how to determine what is an acceptable level of risk/safety for the public.[10] By analyzing the low-level radiation case, it will be possible to gain further insight on how to resolve these same questions regarding other problematic technologies. The importance of such an assessment ought not to be underestimated; for example, according to Robert A. Roland, president of the Manufacturing Chemists Association, "whether government understands, accepts, and applies risk/benefit analysis to regulation will be the most consequential question facing the chemical industry in the 1980's".[11]

Regardless of whether one applies risk/benefit analysis to the chemical industry or the nuclear industry, the potential ethical significance is the same. This is because, in both instances, whoever affirms or denies the desirability of current chemical or radiation standards is, to some degree, symbolically assenting to a number of American value patterns and cultural norms. In advocating present radiation policy one might also be supporting, for example, an ethic of economic progress/growth, rather than one of no-growth; a 'realistic' morality, rather than a romantic or utopian one; a utilitarian system of values, rather than an egalitarian one;[12] a social ethic based primarily on industrialization, rather than on environmental integrity as well; a political ethic guaranteeing the rights of a regulatory elite, rather than those of all citizens; or a morality based on maximizing military and industrial strength rather than citizens' health and safety. In other words, the policy dispute regarding radiation standards is far more than a conflict over scientific opinion. It is also a symbolic vehicle in terms of which implicit and far-reaching ethical assumptions are made explicit, and thus susceptible to evaluation, by the members of a culture.

In attempting to promote such an assessment of cultural norms, this chapter will focus on two tasks. The first is to illuminate some basic American value patterns by analyzing radiation standards undergirding regulation of nuclear technology. The second task is to engage in a carefully reasoned moral criticism of present public policies regarding low-level radiation. To accomplish both of these aims, the discussion will proceed in terms of three basic steps. First, there will be an outline of current federal regulations governing low-level radiation. Secondly, four questionable assumptions implicit in this public policy will be uncovered. These presuppositions are: that, in allowing radiation releases to the environment, economic benefits outweigh health costs; that whatever is normal is moral; that persons most likely to benefit from radioactive emissions are those who ought to be charged with controlling them; and that the states ought not to set radiation standards more stringent than those of the federal government. After arguing that all four of these assumptions are suspect on both logical and ethical grounds, the chapter will consider three undesirable consequences of current public policy governing low-level radiation. These results are: that equity is not served; that rights to equal protection and to due process are violated; and that a higher priority is given to technological development than to public health/safety or to local control of policy affecting local hazards. On the basis of the logical and ethical problems implicit in these assumptions and consequences, the chapter concludes that American public policy regarding nuclear technology exhibits a dangerous trend. Let us now examine precisely why this conclusion follows.

2. Federal Radiation Standards

The most basic federal requirement governing low-level radiation is that emissions from a nuclear power plant are not allowed to cause any member of the public to receive more than a 0.5 rem whole-body radiation dose in any calendar year.[13] Although 0.5 rem is the maximum permissible dose, reactor licensees are required to "make every reasonable effort to maintain radiation exposures. . . to unrestricted areas, as low as is reasonably achievable".[14] What is "as low as is reasonably achievable" is fixed on the basis of a "favorable cost-benefit analysis".[15] If it costs the licensee more than $ 1000 to avoid an additional man-rem of exposure to the public, then he is not required to do so.[16] So long as this cost is under $ 1000 per man-rem, the licensee must aim at reducing maximum radiation exposure to the public to 0.0005 rem per person per year.[17]

Implementation of the radiation standards just outlined is a-chieved by means of quarterly emission reports sent by the power plant licensee to the Nuclear Regulatory Commission. In the event that any member of the public is exposed to radiation from a plant in excess of "ten times any applicable limit", the licensee is required to notify the NRC within 30 days. For exposures to the public in excess of 500 times any limit, notification within 24 hours is required. Finally, for exposures in excess of 5000 times any limit, the licensee is required to make immediate notification to the NRC.[18] "Any person who willfully violates any provision" specified in the Code is subject to fine, imprisonment, or both.[19]

Although intentional failure to follow maximum exposure guidelines is punishable by law, the Code also provides that the NRC "may. . . grant such exemptions from the requirements as will not result in undue hazard to life or property". Such exemptions "may include proposed radioactivity limits higher than those specified. . .if. . . the applicant has made a reasonable effort to minimize the radioactivity contained in effluents to unrestricted areas". However, no city or state has the power to set more stringent emission requirements. "The Federal government has exclusive authority. . . to regulate the discharge of radioactive effluents."[20]

3. Ethical Problems of Radiation Policy

On the basis of this brief summary of regulations governing low-level radiation, one is able to discover a number of important assumptions implicit in public policy. Perhaps the most basic ethical presupposition in current radiation regulations is that the economic and technological benefits gained by permitting some radioactivity to be dispersed in the environment are worth an increase in cancer and genetic damage. In assuming that this tradeoff is indeed a moral one, the authors of the current policy are clearly following a utilitarian, rather than an egalitarian, ethic. In other words, they are guided by a principle of maximizing the greatest amount of good for the greatest number of people, rather than by one of guaranteeing equal justice for all.

3.1. ADHERENCE TO THE PRINCIPLE OF UTILITY

According to the utilitarian account of moral obligation, the sole basic standard of right and wrong action is the 'principle of utility'. It posits, as the moral goal of all human actions, the greatest possible balance of good over bad for mankind as a whole.[21] Since the moral end of utilitarianism is to maximize good, principles of individual rights and equal justice are recognized only to the extent that doing so will lead to the greatest good. As a consequence, application of the tenets of utilitarianism can lead to individual violations of equity and justice, at the same time as the good is maximized for mankind as a whole.[22] For example, says J. J. C. Smart, a utilitarian would be bound to accept as right the action of framing and executing an innocent man, if his death would prevent serious riots which would take the lives of many persons.[23]

According to (what I have called) an 'egalitarian' ethic, on the other hand, the moral goal is not maximizing good for all mankind. Instead, egalitarians follow a principle based on "equality in the assignment of basic rights and duties", apart from the totality of good achieved by such a principle.[24]

Clearly, if current public policy leads to the consequence that some persons bear a higher carcinogenic and mutagenic risk than others (because of their exposure to radiation), then that policy is not egalitarian but utilitarian. Even if one assumes that the technological and economic benefits arising from allowing such exposures lead to the greatest good for the greatest number, the policy of permitting these exposures is problematic. The difficulty with this and with any utilitarian policy is that it allows, in principle, disenfranchising minorities of their rights in order to serve the good of the majority. But if minorities may be divested of rights under a utilitarian framework, then the very concepts of equal justice and of individual and inalienable rights become meaningless.

Utilitarians recognize this consequence of their position. As Smart points out, following the principle of utility often leads to maximizing the good of mankind as a whole, while avoiding sacrificing individual rights to the common good. Nevertheless, he admits, "it is not difficult to show that utilitarianism could, in certain exceptional circumstances, have some very horrible consequences".[25]

In response to this criticism, that within a utilitarian framework minorities might be disenfranchised of rights in order to serve the alleged good of the majority, utilitarians maintain that such violations of equity or justice are "the lesser of two evils (in terms of human happiness and misery)".[26] In other words, they claim that more human suffering might be caused by following principles of equity than by attempting to maximize the good of all people. Another utilitarian response to this criticism is that the right to equality of treatment is not absolute; to claim otherwise, they say, would delay 'social improvement'.[27]

The problem with both these utilitarian answers, however, is that they make ethics a matter of expediency. The whole point, of guaranteeing equal justice and inalienable rights, is to protect every individual from capricious or expedient denials of these rights. But if, by the utilitarian definition of morality, one is bound to maximize the good of mankind as a whole, apart from whether individual rights are served, then it is meaningless to posit inalienable rights to equity and justice. If these principles are served only when they do not conflict with 'the greatest good for the greatest number', then they are recognized only when it is expedient to do so, rather than because humans have inalienable rights to them.

The purely rational problem, with inequitable distribution of technological costs and benefits, is argued well by John Rawls. If one were ignorant of his natural ability, place in society, and other factors that might influence his share of society's goods, says Rawls, then one would be able to assent to principles of justice that were the consequence of fair agreement or bargain, rather than the result of personal hope for advantage. In such a situation, he states, no rational person would agree to principles of utility, that would sanction "an enduring loss for himself in order to bring about a greater net balance of satisfaction". Instead, individuals would choose principles of "equality in the assignment of basic rights and duties". They would sanction inequalities in wealth and authority only to the extent that "they result in compensating benefits for everyone, and in particular for the least advantaged members of society".[28]

Inequities do not merely affront one's rational sensibilities, however. They are also incompatible with the principles upon which the US Constitution rests. As Marshall Cohen, professor of philosophy at The City University of New York, has pointed out, for some time the dominant US moral philosophy has been utilitarian in its assumptions. It has wrongly been the basis for urging minorities to submit to the interests of the majority. But, as Cohen argues,

utilitarian attitudes are incompatible with our moral judgment and with the principles on which our Constitution rests. It is, therefore, a crucial task of moral and political philosophy to make clear the inadequacy of utilitarian concepts and, more important, to provide a persuasive alternative to them.[29]

In arguing against utilitarian standards for low-level radiation, one is really arguing for recognition of the constitutional rights of all persons. Equal protection is one such right, guaranteed by the Fifth and Fourteenth Amendments. Since its recognition is not contingent on whether it is economically beneficial to do so, these rights are said to be inalienable. When conditions (e.g., promoting economic or technological well-being) are put on the exercise of these rights, then they can be said neither to be guaranteed by the Constitution nor to be inalienable. Hence utilitarian theories are not compatible with the principle of equal rights upon which our Constitution rests.

3.2. VIOLATIONS OF EQUAL RIGHTS

Some of the minorities, whose equal claims to rights are compromised by current radiation standards, include children and citizens living within a 50-mile radius of a nuclear plant. Other classes of persons whose equal rights are not protected include those who are susceptible to the risk of cancer by virtue of previous medical exposures to radiation, and members of future generations who will die as a result of radiation-induced genetic damages.

The case of children presents a particularly instructive example of how the radiation burden is not borne with equal justice. If a child and an adult each receive one rad of total body radiation, the child (according to government calculations) is three to six times more likely than the adult to contract cancer because of this exposure.[30] This means that if children receive the maximum permissible annual dose of radiation for only two years, then they are *prima facie* exposed to unequal risks and unequal protection. These consequences assume an even greater magnitude when one realizes that radiation exposure is not short-lived, but cumulative, and that the average lifetime of a nuclear plant is 30 years.

Persons living within a 50-mile radius of a reactor also constitute a minority whose rights to equal protection are compromised. Public policy makers clearly admit that, because of their proximity to a nuclear plant, these people bear a cancer and genetic damage risk up to as much as fifty times greater than that borne by the general US public.[31] Moreover, since it is generally admitted that any amount of radiation increases the probability of cancer and genetic damage, reactor emissions will also cause a much greater health risk to be borne by those who have already received high doses of radiation, e.g., through medical X-rays. Hence once again, by applying equal radiation exposure limits to all persons, one is engaging in discrimination.

This same charge is also true relative to future generations. According to government estimates, every rad of radiation causes 0.002 genetic deaths among offspring of irradiated ancestors.[32] If, in the thirty-year lifetime of a particular reactor, all the people in a given area annually receive the maximum permissible annual dose of low-level radiation (0.5 rem) from the plant, then at least 3% of these people will produce children who will die from genetic disorders induced by their 30-year exposure to additional radiation.[33] This means that 3% of the future population in a given area might be deprived of equal protection for the sake of the majority who are allegedly benefitted by current radiation standards and consequent utility rates and energy production. Even if one is willing to grant that human life and health ought to be traded for technological benefits, there is still the problem that the liabilities of such a tradeoff are not borne equitably. As was argued previously, to the extent that our cultural norms follow the Fifth and Fourteenth Amendments and are based on concepts of equal justice, equal protection, and due process for all, these utilitarian radiation standards run counter to long-standing American value patterns.

3.3. CONFUSING WHAT IS NORMAL WITH WHAT IS MORAL

Public policy regarding radiation also symbolizes an American emphasis on utility by virtue of its monetary criterion for pollution control. The policy explicitly postulates that, once the 0.5 rad annual standard is met, it is worth $ 1000 (but no more) to prevent genetic damage and cancer arising from one man-rem of whole-body radiation.[34] Even if one agrees with government data regarding the low estimates of death and damage induced per rad of exposure, and with the necessity of fixing a price on human life, there might still be reason to question why all persons are not given equal dollar value and equal protection. One might also think of the $ 1000 criterion, not only as the price paid for certain numbers of deaths and diseases, but also as the cost affixed to the questionable privilege of injuring our greatest treasure, our genetic inheritance.[35]

The chief methodological assumption, on which the 0.5 limit and the $ 1000 criterion are built, is that if humans have lived with x amount of natural radiation with only 'negligible' consequences, then it is moral to permit technology to add to this level by a factor of two or less, since these consequences are also likely to be negligible.[36] Using this same assumption, one might reason that, if x numbers of cars 'naturally' operate with faulty brakes and are responsible for a negligible number of automobile accidents, then it would be moral to allow $3x$ total numbers of cars to operate with faulty brakes, on the supposition that their effect would also be negligible. False as this assumption is, it is at the heart of all 'naturalistic' ethics; it rests on the proposition that whatever is 'normal' (e.g., cancers induced by background radiation) is moral.[37] As G. E. Moore was careful to point out, however, the normal is of necessity neither good nor bad, and hence is not always moral.[38] Because current radiation standards are built on the assumption that what is normal is moral, they are an apt symbol both for society's willingness to view the status quo as good, and for its unwillingness to try 'abnormal' ways of thinking, producing energy, and running an economy. For this reason, radiation standards bear obvious testimony to the fact that society has sanctioned the normality (and therefore the morality) of a high-technology, pollution-tolerant style of life rather than an 'abnormal' one built around soft or alternative technologies.

One paradoxical aspect of defining man-made radiation as normal and therefore moral, because it is of the same degree of magnitude as background levels, is that this stance is inconsistent with current US public policy regarding fallout. The US stopped above-ground testing of nuclear devices, in part, because of its policy of not polluting the atmosphere with radiation. Yet reliable government documents reveal that global fallout adds only 0.004 rem/year of whole body exposure, while current atomic plant radiation standards permit 0.5 rem/year of whole body exposure.[39] At the least, our radiation standards are a symbol of the apparent incompatibility of various policies regarding nuclear technology.

3.4. FAILURE TO OBTAIN DISINTERESTED MONITORING

Another surprising aspect of the same public policy is that nuclear licensees, the persons most likely to cause excessive radiation and to profit from it, are those entrusted with monitoring it.[40] This practice clearly violates principles of fair play and disinterestedness. Equally obvious evidence, for the extent to which radiation monitoring symbolizes the triumph of special interests, is the fact that even gross violations of radiation standards need not be reported immediately to the NRC. If a radiation emission is in excess of 500 times the 0.5 rem limit, for example, the licensee is required to notify the NRC only within 24 hours,[41] even though at least half the persons exposed would be likely to produce offspring who would die from genetic damage.[42] Since the incidents are not reported immediately and since there is no independent check on radiation levels, there is no accurate way of knowing whether radiation was 500 or 5000 or 50,000 times the acceptable limit at the time of an accident. There also would be no way to determine that consequent genetic deaths were caused by this incident and not by some other factor; there are no provisions for mandatory follow-up studies after violations of exposure limits, and even if there were, statistical correlation is not sufficient to prove particular causality of cancer or genetic damage. This means that current radiation policy is likely to lead not only to violations of equal protection but also to a situation in which, in principle, these violations cannot be proved.

If the NRC really means to guarantee to the public that radiation standards will be met, then its policy of allowing reactors to continue to operate, after standards have been exceeded by a factor of 500, is puzzling. If radiation standards are public policy and not public pacifiers, then at a minimum this policy ought to include daily emission monitoring by a disinterested person,[43] and immediate shutdown of the reactor so long as exposure limits cannot be met. Secondly, if government intends to protect the public against radiation hazards, then (contrary to current public policy) all violations of permissible radiation levels ought to result in application of some negative sanction to the violator, as well as compensation to those persons who have been irradiated. The sanction and compensation ought not to depend, as they now do, on proving that the violations were 'willful' and that they have resulted in observable health effects. As is well-known, many radiation-induced cancers have latency periods of up to 40 years. Hence it is unlikely that most carcinogenic effects would be immediately observable. Moreover, whether they are caused willfully or not, all violations of radiation standards are harmful, and hence all deprive citizens of equal protection for which they deserve compensation.[44]

3.5. CONFUSING OBJECTIVE AND SUBJECTIVE MORALITY

Legally-permitted violations of current radiation standards also err in being inconsistent with accepted scientific fact and in confusing objective and subjective morality. The government cannot claim, both that any amount of radiation is hazardous (as it does, correctly), but that the NRC may grant 'exceptions' to standards, provided that the result will not constitute an 'undue' hazard to the public.[45] If any amount of radiation is both hazardous and a violation of equal protection and equal justice, then how can a legally-permitted increase in radiation not be an 'undue' hazard? Moreover, since standards need not be met, so long as the NRC judges that the licensee shows 'a reasonable effort' at meeting them, current policy allows government regulators to trade human health and welfare for the good intentions of the promoters of technology.

Good intentions have never been shown to constitute sufficient conditions for the morality of an act. Hence, to say that an *objective* ethical *obligation* (to meet radiation standards) has been fulfilled, so long as one has the *subjective* ethical *intention* to do so, is clearly incorrect. It confuses the rightness/wrongness of *acts* with the culpability/inculpability of *agents*. Although good intentions are often sufficient to remove the culpability of an *agent*, they are not sufficient to render an objectively wrong *act* thereby right. Moreover, if one assumes that good intentions constitute sufficient conditions for the morality of an act, then many notoriously wrong acts may be said to be objectively right. (Witch hunts, cross burnings, and crusades against Infidels, Jews, and leftists, for example, have all been sanctioned at one time or another by those who believed that the good intentions of their perpetrators were sufficient grounds for their acts to be considered moral.) For both these reasons, it is unlikely that one ought to assume that good intentions constitute sufficient conditions for the morality of an act.[46] But if they do not, then one ought not to assume that the intention to meet radiation standards is a sufficient condition for affirming that failure to meet such standards is objectively morally acceptable. But if this thesis ought not be assumed, and current public policy is built upon it, then this radiation policy is also suspect.

3.6. INCONSISTENT APPLICATION OF FEDERAL PREEMPTION

Although numerous other ethical presuppositions of current

radiation policy ought to be examined, there is one issue which, perhaps more than any other, raises the symbolic dimensions of an important conflict in American values. This controversy, over federal preemption of state and local rights to control radio-active emissions, is an important focal point both for numerous debates over states' rights and for the tension between public order and private liberty. The crux of the dispute, preemption, is a doctrine resting on the supremacy clause of the federal con-stitution. It provides that the constitution and the laws of the US, rather than state laws, shall be supreme in the land. Applied to public policy regarding radiation, preemption requires that all state and local governments hand over authority for setting radiation standards to the NRC as the appropriate federal regu-lator.

On the one hand, opponents of nuclear technology argue that forcing a state to accept hazardous emissions of power reactors is a clear violation of the state's authority. They claim that, because all other pollution control is based on the premise that the federal government ought only to look after minimum safety standards, while the states are always free to "develop standards more stringent than the federal criteria", current public policy regarding radiation limits is unjust. It is unfair, they maintain, because numerous states have not been allowed to set standards which are stricter than those approved by the NRC,[47] even though protecting the public health and safety has always been a responsi-bility of states.[48]

On the other hand, proponents of nuclear technology argue that, by virtue of the 1954 Atomic Energy Act, the federal government is required to promote the peaceful uses of atomic energy. If states were allowed to set tighter radiation standards for plants within their boundaries, they would be permitted to hinder development of atomic energy. Moreover, they claim, Section 274 of the Atomic Energy Act "constitutes an express congressional declaration that radiation hazards ... are to be within the exclusive jurisdiction of the NRC".[49]

Current public policy dictates, by virtue of a recent Supreme Court decision,[50] that only the federal government (through the NRC) may regulate radioactive emissions from a power plant. Historically, however, states have always had responsibility for protecting the public health and safety; whenever state laws have been designed to provide better protection in this area, they have not been preempted.[51] Moreover, (beginning prior to development of nuclear power plants) states have exercised regulatory authority over various radioactive materials, e.g., those produced in research facilities, naturally occurring, or released by X-ray equipment.[52] In all matters involving nonradioactive environmental pollution, states have always had the right (and have been encouraged) to develop standards more stringent than federal guidelines.[53] State control probably has been championed because it is thought to provide a number of advantages. Some of these benefits include: a brake on the acceleration towards more centralized government;[54] a check on the tendency for a technocratic elite, rather than the people, to make local policy;[55] and a framework within which public policy is made by those who are most likely to bear its liabilities, as well as by those who will profit from its assets.[56]

Why has a similar rationale not been employed so that states might also have the right to set more stringent radiation standards? When the Atomic Energy Act was amended in 1959, Congress explained that since the states did not have the necessary radiological expertise, greater protection for the people could be afforded by federal regulation. In fact, "one of the express purposes" of this amendment "was to recognize that, as the states improved their capabilities to regulate nuclear power effectively, Congress should give the states a greater role".[57] Yet when states did develop their technical base of expertise and demanded better protection for their citizens, the courts did not follow the congressional directive to "give the states a greater role". Even though Congress had decided earlier that the question of state versus federal regulation of radiation ought to be decided on the basis of the best means of protecting the public, the NRC and utility lawyers were able to convince the courts not to decide the issue on this basis. They argued that allowing states to set stricter standards would "unnecessarily stultify the industrial development and use of atomic energy for the production of electric power". Hence the courts decided in favor of federal control of radiation standards.[58] The judges feared that, if preemption were not employed in this case, the "total defeat of the valid Congressional plan [for peaceful use of atomic energy, as specified through the Atomic Energy Act] to promote commerce would be the result".[59]

What is puzzling about the courts' reasoning, however, is that nuclear energy is not the only type of power authorized for development by Congress.[60] Moreover, no federal legislation requires every or any state to employ nuclear power.[61] Hence, if states may legally develop and use non-nuclear sources of power, then it is unclear how one can validly argue that state attempts to restrict plant emissions are in violation of the 1954 Atomic Energy Act, whose purpose is to promote use of nuclear energy. On these grounds, it seems equally plausible that those states developing nuclear power (and bypassing other energy sources) might likewise be said to be in violation of a Congressional mandate, viz., the Energy Reorganization Act of 1974, which requires that all forms of energy be developed.

Whether or not states ought to have the right to set more stringent radiation standards, the preemption conflict reveals several trends in American value patterns. First, since the Atomic Energy Act has been used to preempt state control of radiation, even though states are able to set stricter standards for non-radioactive pollution, an interesting disparity emerges. By virtue of recognizing the right of states to set higher criteria for protecting public health and safety, policy regarding non-radioactive pollution allows for the possibility that a higher priority will be given to public health and safety than to technological development. On the contrary, because the Atomic Energy Act has been used to deny states the right to set stricter standards in order to protect the public against radiation, policy regarding this type of pollution apparently demands that a higher priority be given to technological development than to protecting public health and safety.

Likewise, since only radiation-related health and safety statutes are preempted by the federal government, a lower priority is given to local public, than to federal agency, control of policy regarding low-level radiation. On the contrary, policy governing all other health and safety issues reflects the higher priority being given to local public control. Hence policy regarding low-level radiation reflects less concern, both for public health and safety and for states' rights, than does policy regarding other forms of environmental pollution.

4. Conclusion

If the preceding analysis is correct, then American public policy regarding nuclear technology shows a dangerous trend. Current regulation of low-level radiation symbolizes the extent to which our culture has exchanged an ethic of equal justice and equal protection for one of utility. It also indicates the low monetary and moral value placed on human sickness and death when it is the price paid for the benefits of abundant energy. It is significant that even fair play and logical consistency have been bent to serve the needs of a powerful and capital-intensive technology. Moreover, if the previous evaluation of the preemption issue is correct, then radiation policy clearly symbolizes how states' rights, as well as public health and safety, can be legally forced to take second place to technological development. Fortunately, however, there are many Americans who are convinced that the legality of public policy is not synonymous with its morality.

Notes

[1] Cited by M. Korchmar, 'Radiation Hearings Uncover Dust', *Critical Mass Journal* **3**(12), (March 1978), 5.

[2] US Atomic Energy Commission, *Comparative Risk-Cost-Benefit Study of Alternative Sources of Electrical Energy* (WASH-1224), US Government Printing Office, Washington, D.C., December 1974, p. 3-95; hereafter cited as: 'WASH-1224'. According to the *Code of Federal Regulations*, 10, Part 20, US Government Printing Office, Washington, D.C., 1978, p. 184 (hereafter cited as: '10 CFR 20'), one 'rad' of radiation is "a measure of the dose of any ionizing radiation to body tissues in terms of the energy absorbed per unit mass of the tissue".

[3] Arthur R. Tamplin and John W. Gofman, *'Population Control' Through Nuclear Pollution*, Nelson-Hall, Chicago, 1970, pp. 4, 85; hereafter cited as: *Nuclear Pollution*. Damage estimates are made by plotting the curve obtained when various health effects ('response') are interpreted as a function of radiation dose. Since dose-response coefficients for low-level radiation are obtained by hypothesis from data on high levels of exposure (i.e., Hiroshima and Nagasaki), there is no general agreement regarding health effects of low-level radiation. Government scientists assume that the dose-response curve follows a linear pattern according to which there is a small, constant risk per rad, independent of dose rate. Many prominent academicians such as Gofman and Tamplin, on the other hand, believe that there is a rapid rise in the dose-response function (such that there are great damages even from low radiation exposures) followed by a leveling of the curve at higher exposures.

[4] US Environmental Protection Agency, 'Criteria for Radioactive Wastes', *Federal Register* **43**(221), (November 15, 1978), 53266; hereafter cited as: 'Radioactive'.

[5] Tamplin and Gofman, *Nuclear Pollution*, p. 51.

[6] For discussion of the role of scientific judgments versus value judgments in setting standards regarding this NRC report, see William W. Lowrance, *Of Acceptable Risk: Science and the Determination of Safety*, William Kaufmann, Los Altos, California, 1976, pp. 41–44; hereafter cited as: *Acceptable Risk*.

[7] Lowrance, *Acceptable Risk*, p. 114; see pp. 109–114 for more discussion on matters of fact versus matters of value.

[8] Lowrance, *Acceptable Risk*, p. 136.

[9] J.E. Martin, US EPA, *Considerations of Environmental Protection Criteria*

for Radioactive Waste, Government Printing Office, Washington, D.C., February 1978, p. 17.

[10] For an important discussion of these problems, see William Lowrance, *Acceptable Risk*, esp. pp. 4–44, 70–75. On a more popular level, see Fred Hapgood, 'Risk-Benefit Analysis: Putting a Price on Life', *The Atlantic* **243** (1), (January 1979), 33–38.

[11] R.A. Roland, quoted by L.J. Carter, 'An Industry Study of TSCA: How To Achieve Credibility?', *Science* 203(4377), (19 January 1979), 247. I am grateful to Professors Robert E. Snow and David E. Wright, both of Michigan State University, for making me aware of the Carter essay.

[12] Utilitarian versus egalitarian ethics are discussed in detail later in the chapter. See Sections 3.1 and 3.2.

[13] '10 CFR 20', p. 189. According to '10 CFR 20', p. 184, "the rem . . . is a measure of the dose of any ionizing radiation to body tissues in terms of its estimated biological effect relative to a dose of one roentgen of X-rays". For most purposes, one rem = one rad (see note 2) of radiation, since one rem of X-rays or gamma rays = one rad of X-rays, gamma rays, or beta rays. As also defined in '10 CFR 20', p. 184, 'dose' is "the quantity of radiation absorbed, per unit of mass, by the [whole] body or by any portion of the body". Although the 0.5 standard is a 'maximum' one, i.e., set for "all pathways of radiation", it does not include exposures through food chains; these are not required, by law, to be taken into account, although they "may be evaluated at the locations where the food pathways actually exist" (10 CFR 50', Appendix I, p. 373).

[14] '10 CFR 20', p. 182.

[15] Nuclear Regulatory Commission, *Issuances* 5, Book 2, US Government Printing Office, Washington, D.C., June 30, 1977, p. 928; hereafter cited as: *Issuances*.

[16] *Issuances*, p. 980. One 'man-rem' of radiation is equivalent to one person's exposure to one rem of radiation; two man-rems are equivalent to one person's exposure to two rems or to two persons' exposures to one rem, and so on.

[17] '10 CFR 50', Appendix I, p. 372.

[18] '10 CFR 20', p. 199.

[19] '10 CFR 20', pp. 202ff.

[20] '10 CFR 20', pp. 201, 189. *Issuances*, pp. 964, 970.

[21] For discussion of the principle of utility and the utilitarian framework for ethics, see John Stuart Mill, *Utilitarianism, Liberty, and Representative Government*, E. P. Dutton and Company, New York, 1910, esp. pp. 6–24; hereafter cited as: *Utilitarianism*. See also Jeremy Bentham, *The Utilitarians: An Introduction to the Principles of Morals and Legislation*, Doubleday, Garden City, New York, 1961, esp. pp. 17–22; hereafter cited as: *Principles*.

See P. Nowell-Smith, *Ethics*, Penguin, Baltimore, 1954, p. 34, and J.J.C.
Smart, 'An Outline of a System of Utilitarian Ethics', in *Utilitarianism: For
And Against* (ed. by J.C.C. Smart and B. Williams), Cambridge University
Press, Cambridge, 1973, pp. 3–74; hereafter cited as: *Utilitarianism.*
[22] Smart, *Utilitarianism*, pp. 69–71; Mill, *Utilitarianism*, pp. 38–60.
[23] Smart, *Utilitarianism*, pp. 69–70.
[24] John Rawls, *A Theory of Justice*, Harvard University Press, Cambridge,
1971, pp. 14–15. See also Charles Fried, *An Anatomy of Values*, Harvard
University Press, Cambridge, 1970, pp. 42–43; Charles Fried, *Right and
Wrong*, Harvard University Press, Cambridge, 1978, pp. 116–17, 126–27;
and Alan Donagan, *The Theory of Morality*, University Press, Chicago, 1977,
pp. 221–39.
[25] *Utilitarianism*, p. 69.
[26] Smart, *Utilitarianism*, p. 72.
[27] Mill, *Utilitarianism*, pp. 58–59.
[28] Rawls, *A Theory of Justice*, pp. 14–15.
[29] Marshall Cohen, 'The Social Contract Explained and Defended', *The New
York Times Book Review*, (July 16, 1972), 1, 16–17.
[30] AEC, 'WASH-1224', p. 4-14.
[31] AEC, 'WASH-1224', p. 4-16.
[32] AEC, 'WASH-1224', p. 4-14.
[33] (0.002 genetic deaths/rad) x (30 x 0.5 rad) = 3% risk per person, or 3
deaths per 100 exposures.
[34] NRC, *Issuances*, pp. 928, 980.
[35] Gofman and Tamplin view the allowance of any releases of radioactivity
as a 'prescription for genocide' (*Nuclear Pollution*, p. 211).
[36] AEC, 'WASH-1224', p. 4-7 and p. 1-16; see also p. 4-8.
[37] G.E. Moore, *Principia Ethica*, Cambridge University Press, 1951, pp. 39–
60; hereafter cited as: *Principia.*
[38] *Principia*, pp. 58, 43. See also Alasdair MacIntyre, 'Utilitarianism and
Cost-Benefit Analysis,' in *Values in the Electric Power Industry* (ed. by
Kenneth Sayre), University Press, Notre Dame, 1977, p. 217.
[39] AEC, 'WASH-1224', p. 4-9.
[40] '10 CFR 20', p. 199.
[41] '10 CFR 20', p. 199.
[42] AEC, 'WASH-1224', p. 4-14.
[43] Numerous university-based scientists and humanists have made this same
point to both the Atomic Energy Commission and the Nuclear Regulatory
Commission. See, for example, US Atomic Energy Commission, *Nuclear
Power and the Environment*, US Government Printing Office, Washington,
D.C., 1969, pp. 54–55; hereafter cited as: AEC, *Nuclear Power.*
[44] The Atomic Energy Commission, the Nuclear Regulatory Commission,

the US Public Health Service, and the Federal Radiation Council all admit, unequivocally, that any amount of radiation, however small, is biologically and medically harmful to humans. See, for example, AEC, *Nuclear Power*, p. 41; '10 CFR 20', p. 182; and '10 CFR 50'.

[45] See note 20.

[46] For discussion of the distinction between acts/agents, objective morality/ subjective morality, and permissiblility/impermissibility versus culpability/ inculpability, see Donagan, *The Theory of Morality*, p. 37. Here he says: "Common morality, as the Hebrew—Christian tradition understands it, has to do with human actions both objectively, as deeds or things done, and subjectively, as the doings of agents. Objectively, they are either permissible or impermissible; subjectively, either culpable or inculpable".

[47] See, for example, G.B. Karpinski, 'Federal Preemption of State Laws Controlling Nuclear Power', *Georgetown Law Journal* 64(6), (July 1976), 1335, 1341; hereafter cited as: 'Federal Preemption'; and A.W. Murphy and D.B. LaPierre, 'Nuclear "Moratorium" Legislation in the States and the Supremacy Clause: A Case of Express Preemption', *Columbia Law Review* 76(3), (April 1976), 418–419; hereafter cited as: 'Express Preemption'. Murphy and LaPierre explain that in Northern States Power Company v. Minnesota, "the Court of Appeals for the Eighth Circuit held that the federal government has the exclusive authority under the doctrine of preemption to regulate the construction and operation of nuclear power plants and that this authority includes regulation of the levels of radioactive effluents discharged from the plant and precludes such regulation by the states". This opinion was affirmed by the US Supreme Court in 1972; the court ruled that the State of Minnesota could not legally set more stringent standards on emissions from nuclear power plants. Despite this ruling, numerous states have continued to challenge the NRC on the question. In 1975, for example, 24 state legislatures introduced legislation aimed at restricting radiation levels. ('Express Preemption', 392.)

[48] Comptroller General of the US to the Joint Committee on Atomic Energy, 'Opportunities for Improving AEC's Administration of Agreements with States Regulating Users of Radioactive Materials', *Atomic Energy Law Journal* 15(2), (Summer 1973), 65–66, 71; hereafter cited as: 'Comptroller General'.

[49] 'Express Preemption', 407–408, 446, 449; J. R. Brydon, 'Slaying the Nuclear Giants', *Pacific Law Journal* 8(2), (July 1977), 770; hereafter cited as: 'Nuclear Giants'.

[50] See note 47.

[51] This point is made in 'Express Preemption', 434–44; see also 'Comptroller General', 65–66, 71.

[52] 'Nuclear Giants', 764.

[53] 'Federal Preemption', 1341. See also V. McKim, 'Social and Environmental Values in Power Plant Licensing', in *Values* (ed. Sayre), pp. 33–34. M.S. Young, 'A Survey of the Governmental Regulation of Nuclear Power Generation', *Marquette Law Review* **59**(4), (1976), 852–53, also notes that preemption is employed only in radiation-related issues of pollution standards.

[54] See R.B. Stewart, 'Paradoxes of Liberty, Integrity, and Fraternity: The Collective Nature of Environmental Quality . . . ', *Environmental Law* **7**(3), (Spring 1977), 472–73.

[55] See K.M. Rhoades, 'Environmental Law: The Supreme Court Interprets the Role of the Environmental Protection Agency in Regulating Radioactive Materials', *Washburn Law Journal* **16**(2), (Winter 1977), 522.

[56] This point is also made by E.D. Muchnicki, 'The Proper Role of the Public in Nuclear Power Plant Licensing Decisions', *Atomic Energy Law Journal* **15**(2), (Spring 1973), 51.

[57] This Congressional background is provided by Karpinski in 'Federal Preemption", 1340; N. Notis-McConarty, 'Federal Accountability: Delegation of Responsibility by HUD Under NEPA', *Environmental Affairs* **5**(1), (Winter 1976), 136–37; and in Comptroller General, 66–67.

[58] 'Nuclear Giants', 770.

[59] 'Express Preemption', 450.

[60] According to the Energy Reorganization Act of 1974, Congress authorized the development of all sources of energy; see 'Federal Preemption', 1337.

[61] 'Federal Preemption', 1336.

Chapter Three

Nuclear Wastes and the Argument from Ignorance

Senator Howard H. Baker, a member of the Joint Committee on Atomic Energy, said in 1973 that "the containment and storage of radioactive wastes is the greatest single responsibility ever consciously undertaken by man".[1] Despite the gravity of this responsibility, however, there is no National Radioactive Waste Policy,[2] even though the US has been generating the waste for approximately 40 years.

In April 1977, President Carter revealed his determination to establish a nuclear waste program. He created the Interagency Review Group and charged it with developing a method for dealing with the problem of radioactive waste.[3] When the Group reported back to the President in October 1978, they indicated that the first model for a high-level nuclear waste storage facility could be expected to operate initially sometime between 1992 and 1995. Prior to that operation, said the IRG, the Department of Energy would determine both a repository site and a feasible method for storing the waste.[4]

The public policy questions raised by the problem of nuclear waste are substantial ones. According to the US Environmental Protection Agency, there is widespread "public debate regarding the desirability of establishing a moratorium on the production of additional nuclear wastes until such time as a National Radioactive Waste Policy has been adopted", presumably around 1992–1995.[5] What is at issue is the current practice of continuing to generate nuclear wastes when the goverment has failed to evaluate all of the economic and social costs of its storage. In this chapter, I will present an ·epistemological and ethical analysis of some of the problems inherent in the failure to deal with the costs of waste management. First, I will outline the societal price paid for generating and storing unwanted radioactive materials.

1. The Social and Economic Costs of Storing Radioactive Wastes

One of the most important aspects of the nuclear fuel cycle is the management of radioactive residues. The wastes pose a significant problem, in large part because of their extremely hazardous nature. In the US alone, there are in operation approximately 65 commercial nuclear fission reactors, not including another 70 under construction. Each of these plants adds (or will add) approximately 30 tons of radioactive waste per year to already existing amounts requiring storage and/or management.[6] Moreover, the decay times of radioactive elements produced by nuclear generation of electricity are "millions of years for certain of the actinide elements and long-lived fission products"; before this decay is complete, radioactive wastes have the potential for causing death, numerous types of disease, and carcinogenic and mutagenic injury.[7]

Although several writers distinguish between high-level and low-level radioactive wastes, this discussion will not rely on any such distinction, since it "does not necessarily provide a key to potential [radiation] hazards".[8] The distinction between high and low-level wastes is based solely on whether the radioactive residues are from the first cycle solvent extraction system for reprocessing irradiated reactor fuels. If they are, they are called high-level wastes; otherwise they are low-level wastes. Since the origin of wastes (according to which they are classified as high or low) provides no complete information on their harmfulness, the distinction is not always a useful one. Certain low-level disposal facilities have accepted "substantial quantities of intensely radioactive wastes including strontium 90, cobalt 60 and cesium 137 and a significant amount of plutonium".[9] Since low-level waste repositories contain long-lived radioactive isotopes such as these, having "potential dose commitments for hundreds of thousands to millions of years",[10] they may be said to pose a hazard as significant as high-level waste repositories. "*Transuranic* contaminated wastes (isotopes with atomic number greater than 92, e.g., Np, Pu, Am) are especially significant because they have long half-lives and . . . are persistent in the environment."[11]

Because of the extremely hazardous nature of even small amounts of radioactive wastes (e.g., a millionth of a gram of plutonium can cause lung cancer), it is imperative that they be isolated completely from the biosphere for extremely long periods of time.[12] If they were not properly stored, the current US inventories of nuclear wastes would be enough to kill several billion people.[13]

Owing to the dangers posed by radioactive residues produced during the nuclear fuel cycle, one would think that the social and economic costs of storing them would be included within cost-benefit analyses of nuclear generation of electricity. This, however, has not been the case. In the US, for example, there are a number of reasons why costs associated with managing radioactive wastes and compensating for their health effects have not been accounted for within evaluations of atomic power. For one thing, it is difficult to quantify many items having "significant impacts on human health and safety".[14] Although it is known, for example, that every rad of radiation causes approximately 0.002 genetic deaths among offspring of irradiated ancestors,[15] specific causes of mutations are difficult to identify; it is also a problem to separate mutagenic injuries caused by chemicals, or by background and medical exposures to radiation, from those caused

by accidental releases of stored radioactive wastes from the nuclear fuel cyle.[16]

Secondly, although it is known that radioactive wastes must
be kept in essentially permanent isolation from the biosphere, the
cost of doing so has not been addressed and included in economic
analyses because it is not known. The reason why this price is not
known is that, although various interim methods have been employed, no final storage techniques for high-level radioactive
wastes have been determined or adopted.

At least in the US, all existing means of managing these wastes
have resulted in major leaks of radioactivity into the biosphere.
Plutonium, for example, has travelled off-site from both high and
low-level storage facilities.[17] At the largest commercial storage
site for high-level waste, located in Hanford, Washington, over
500,000 gallons of high-level waste have leaked accidentally from
storage containers.[18] US EPA officials indicate that, because of
migration patterns of radioactivity released directly to the soil
and water at Hanford, normal annual radiation releases from this
facility "could result in a yearly impact of 580 man-rem total
body exposure".[19] In fact, they say, more than 50% of the radioactivity released directly to the soil at certain Hanford sites reach-
es the Columbia River (via groundwater) in four to ten days.[20]
Based on the great number of radioactive leaks from high-level
storage tanks, the government has stated: "extrapolation of past
data would indicate that future leaks may occur at a rate of 2 to
3 per year".[20] Releases of radiation have also occurred at the two
other major waste disposal areas in the US (Savannah River and
Idaho National Laboratory), including widespread plutonium
contamination in the ground water.[22]

Because of all these problems with radioactive leaks, the US is
searching for alternative means of waste containment. The latest
government plan calls for "initial operation of the first high-level
waste disposal facility" in 1992–1995.[23] Moreover, even if a

storage scheme were available now, or had been available before the first plants began fission generation of electricity three decades ago, the problem (of knowing the social and economic costs of storage) would not be resolved. This is because, although permanent storage or disposal is necessary, experts agree that it is in practice impossible to predict, beyond the next 100 years, what the institutional conditions and costs associated with such storage or management will be or whether storage or management will even be possible.[24]

A fourth reason why these costs have not been included within analyses of the nuclear fuel cycle is that, if they were, then cost-benefit considerations would force utilities not to employ fission generation of electricity, despite the fact that most countries of the world are in the midst of an energy crisis. As one nuclear engineer (who sat on licensing boards for commercial reactors for the US Nuclear Regulatory Commission) has stated: if "the licensing board should have to factor into the cost-benefit balance the absence of any in-place solution for waste disposal . . . then . . . I couldn't license any reactor".[25] As it is, however, the absence of a storage plan has been ignored for 40 years, in part because waste storage is not the economic problem of the utility which is generating nuclear waste. Current US policy is that the government owns (and therefore has responsibility to care for) wastes generated within the US and that, in the near future, it might accept the responsibility of safeguarding the wastes of other nations.[26]

2. Philosophical Errors in Analyses of the Waste Problem

Although there are numerous reasons why producers and consumers of electricity have not included costs related to nuclear waste management within their evaluations of nuclear power, the practical effect of this omission has been to ignore a fundamental "law of ecology", that there is no such thing as 'a free lunch'. Ecologically and economically speaking, it is impossible to get something for nothing. US cost-benefit studies of nuclear energy ignore this principle as they fail explicitly to include costs associated with long-term storage of atomic wastes. In so doing, they assume implicitly that one can obtain a *gain*, viz., nuclear-generated electricity, without also incurring a *loss*, viz., the social, economic, and ecological costs of nuclear waste management. In order to understand why this assumption is implausible, let us look more precisely at US economic evaluations of the costs of various modes of generating electricity. Let us first examine the epistemological underpinnings of these analyses. Specifically, what presuppositions underlie the current practice of failing to take account of the costs of storing nuclear wastes?

2.1. EPISTEMOLOGICAL DIFFICULTIES

The most basic epistemological assumption of this practice is that, since we are in ignorance regarding the social, medical, and economic costs of nuclear waste storage, we are therefore justified both in not including these parameters in cost-benefit analyses of comparative sources of electrical power, and in concluding that nuclear fission is a cheaper source of energy than is coal. This assumption is stated explicitly by US policymakers. After admitting that they have no idea whatsoever of (a) the magnitude of the radioactive impact of nuclear waste disposal facilities on human life, (b) the long-term costs of nuclear waste storage, (c) how much radiation will be released to the environment, and (d) how much air and water will be contaminated as a result, US officials state that,

the large uncertainties in the enviromental, health, and injury figures do not alter the conclusion that these costs are small . . . . None of these shortcomings, however, invalidate the major conclusions and findings of the study . . . [that] coal . . . is the most severe environmental offender . . . [and that] direct environmental impacts . . . are essentially absent in the nuclear fuel cycles.[27]

Moreover, although both the Atomic Energy Commission and the Energy Research and Development Agency concluded that nuclear-generated electricity costs only 75% of the price of that produced by coal,[28] both groups have ignored the cost of permanent "surveillance, storage, and control" of radioactive wastes.[29]

2.1.1. *The Argument from Ignorance*

In assuming that one can draw valid conclusions about the social and economic costs of nuclear power while ignoring the costs of waste storage, government policymakers are really committing the argument from ignorance, the *argumentum ad ignorantiam.* Although arguments invoving such a fallacy never establish logically what they are intended to prove, employment of the fallacy might in this instance be understandable if atomic power were the only available energy option. In such a case, our failure to calculate all the costs of nuclear-generated electricity might not be regarded as an inadmissible lack of completeness in our public policy analyses. However, this is not the situation, at least in the US. Here, nuclear power is currently justified by government policymakers largely on the basis of its favorable cost-benefit comparison with the price of producing electricity from coal. Although such a rationale (for using atomic fission) might not be given in countries lacking great supplies of coal, it is neverthe-less an interesting one from the point of view of US economic policy.

Even though US government economists affirm that coal is the most abundant domestic fuel resource, they claim that nuclear power is cheaper than coal, and that "direct environmental impacts . . . are essentially absent in the nuclear fuel cycles" while coal "is the most severe environmental offender".[30] The critical logical and epistemological fault in drawing these two conclusions is, of course, that a major parameter (waste management) affecting both the costs and the environmental impacts of nuclear energy has not been considered in the analysis. The obvious consequence of ignoring this parameter is methodologically dangerous. In so doing, one runs the risk of supporting public policy based on economic conclusions which could be false, e.g., that coal is a cheaper and less environmentally deleterious source of energy than is nuclear power.

The likelihood of employing false economic conclusions becomes more probable when one realizes that both the cost and the health effects of radioactive storage are likely to be considerable. The US Department of Energy estimates that up to $25 billion will be needed in the next 21 years to store radioactive wastes in this country.[31] The 1979 US waste management budget is $449 million.[32] Using both these figures, it is reasonable to assume that average annual costs for waste storage in the US will be from $449 million to $1.19 billion (or $25 billion divided by 21 years). If this assumption is correct, then the permanent cost of radioactive management is very great. The health effects of such storage are also likely to be considerable, if one remembers that all current waste storage sites have resulted in offsite contamination by long-lived radionuclides.[33] According to government estimates, just the normal radiation releases from only one of approximately a score of waste storage sites in this country could result in an annual exposure of 580 man-rem.[34] This degree of exposure from only one site would cause approximately 12 cases of cancer and 116 genetic deaths over a 100-year period.[35] When one realizes that these exposures could go on for a million years and at a number of sites, the magnitude of deaths and cancers becomes quite high.

2.1.2. *Ignoring Monies Budgeted for US Waste Storage*

Even if one argued that it is impossible to calculate exact figures for the health and financial costs of waste storage, this should lead one either (a) to refrain from drawing conclusions about the costs of nuclear-generated electricity, relative to other energy sources, or (b) to provide approximate figures for the costs of storing radioactive wastes. If one chose option (b), then one would still be able to draw some conditional conclusions regarding the relative economy of various modes of generating electricity, including nuclear fission. As it is, however, US government economists and makers of public policy have followed neither plan (a) nor plan (b), and hence have committed the argument from ignorance.

Simply by using actual and projected US budgetary expenditures for radioactive waste storage, one is able to obtain an approximation to the financial costs of attempting to isolate nuclear wastes permanently and completely from the biosphere. Although this figure is low, since it does not include health costs borne by those exposed to normal and accidental releases of radioactive wastes, it is a first step in an attempt to fix a price on waste management. Let us examine this approach.

If one ignores expenditures for waste storage, then US economists tell us that the 1980 annual operating and maintenance costs for one 1000-MWe coal plant are $ 5,400,000 as compared to $ 5,200,000 for a nuclear plant of the same size.[36] But if approximately half a billion dollars will be spent in 1980 for nuclear waste management by the US government,[37] then it is conceivable that, were this cost prorated among nuclear facilities producing the waste, the annual cost of operating and maintaining a nuclear power plant might be higher than operating and maintaining a coal facility of the same size. If one assumes that only half of the approximately $ 500 million to be spent in 1980 for US nuclear waste management is necessary to care for the radioactive by-products of commercial reactors (as opposed to the money necessary to manage military wastes),[38] then the prorated cost of annual storage might be $ 250 million divided by the number of reactors needing to store waste.

Since anywhere from 65 to 135 commercial fission plants will have been operating in the US by 1980,[39] the annual cost of waste management, per reactor, could be said to be anywhere from $ 3.8 million ($ 250 million divided by 65) to $ 1.9 million ($ 250 million divided by 135). Both these figures are considerably more than the $ 200,000 annual maintenance and operation cost-differential between single nuclear and coal plants of the same size (1000 MWe). Moreover, neither figure includes the health costs to the public associated with waste management. Hence it is plausible to conclude that failure to consider the cost of nuclear waste storage in comparative economic analyses of alternative energy sources, not only is part of an invalid argument from ignorance, but also itself may lead to public policy based on erroneous economic conclusions regarding the relative costs of nuclear generation of electricity.

US government policymakers have conceded that they may not be able to successfully employ nuclear waste storage so as "to provide reasonable assurance of long-term isolation of radionuclides".[40] Because of this admission, there is little logical foundation for their related conclusion that nuclear power presents fewer dangerous environmental effects than does generation of electricity from coal.[41] The incomplete substantiation for such a conclusion is all the more apparent when one considers, for example, that the US Environmental Protection Agency has not even formulated standards for "acceptable levels of radioactivity in the environment" from nuclear waste.[42] Hence, not only is the magnitude of probable environmental damage from nuclear waste unknown, but also the criteria for assessing such risks are unknown. This means that ignoring the environmental cost of nuclear waste storage is epistemologically questionable, because of the argument from ignorance, and methodologically doubtful owing to incompleteness of the economic data used to justify current conclusions regarding the relative costs of coal and nuclear power.

2.1.3. The Consequences of Assuming that Nuclear Electricity Is Inexpensive

One obvious consequence of such an argument from ignorance is that more ecologically desirable energy choices, e.g., solar power, might be ignored because of the apparent low cost of fission generation of electricity. Once enormous capital investments are made in a technology such as nuclear fission, it becomes increasingly difficult to change one's investment priorities and to expend funds on other energy technologies. This resistance to change occurs, in part, because of the hope of obtaining a return on the earlier capital investments. In the last 30 years, almost $ 200 billion has been spent on US government subsidies of nuclear fission technology.[43] Currently 10 to 20 times more US energy funds are spent on nuclear power than on solar energy research and development.[44] Such patterns of expenditures are likely to continue, so long as public policy analyses portray atomic energy (relative to other sources of power) as inexpensive. Thus, if nuclear power is in reality not as inexpensive as has been thought, then ignoring the costs of radioactive waste could have the consequence both of skewing energy choices in the future and of causing policymakers to underestimate the potential of 'soft' technologies such as solar energy.

There is already some evidence, in fact, that ignoring the costs of waste management has contributed to failure to consider alternative energy strategies. In the most recent environmental impact assessment of the largest commercial nuclear waste storage site in the US, Hanford Reservation, alternatives to further production of nuclear waste were not even considered in the analysis of relative costs of environmental impacts resulting from various waste policies. Even though approximately 500,000 gallons of high-level waste have leaked from their storage tanks, and although measurable radiation has traveled far offsite into the Columbia River and even into the Pacific Ocean, the health costs of numerous waste accidents have not been computed and analyzed. Rather, it has been maintained that the environmental effects of nuclear energy are negligible and that atomic power is the cheapest source of electricity available.[45]

Another consequence of portraying fission power as relatively inexpensive is that it is likely to increase our dependance on atomic energy. And, as our dependance on this source of power increases, "our tolerance of inadequate and dangerous waste disposal solutions will also increase".[46] Since there are no complete standards for assessing the safety of current storage practices, the cost of adequate waste management procedures has not been determined. Failure to include this parameter within cost-benefit analyses suggests that it has not been properly studied and regulated and hence that, if it continues to be ignored, health costs to the public are likely to accelerate. For further discussion of this point, let us turn to a consideration of the ethical presuppositions and consequences of economic policies concerning nuclear waste storage.

2.2. ETHICAL DIFFICULTIES

The practice of failing to include the costs of nuclear waste storage in economic analyses of atomic power is built on several questionable ethical assumptions. First of all, this practice runs contrary to expressed US government policy stipulating that the costs of nuclear waste storage should be borne by those who generated it and by those who benefitted from the atomic power produced during its creation.[47]

2.2.1. *Violations of Equity*

The most obvious class of persons who will probably bear higher health and financial risks, because of nuclear waste storage, are members of future generations. If radioactive hazards are long-lived and cumulative, and if some radioactive materials (including long-lived transuranics) are dispersed directly to the soil and water, then this can only mean that our descendants will bear a greater social cost from nuclear fission generation of electricity than we bear now.[48] But if future persons are subject to disproportionate costs, then this clearly violates principles of equity. Current US public policy affirms that present persons should not impose risks on "a future generation . . . greater than those acceptable to the current generation".[49] So long as the costs of radioactive waste storage are not included in economic analyses of nuclear fission, then there is no economically feasible way, within a cost-benefit framework, either to include compensation for those persons who bear a greater risk, or to know whether the energy/equity tradeoff of nuclear power is an ethically desirable one.

Besides members of future generations, there are at least three other classes of persons who bear a disproportionate hazard as a result of government failure to assess the financial and health costs of nuclear waste management. Because radiation levels off-site near a waste storage facility are higher than levels normally occurring in the same area, persons living near such sites currently bear greater risks both of radiation accidents and of low-level exposures. Moreover, because effects of radiation are cumulative, those persons who already have had X-rays for medical reasons also bear a greater risk from possible exposure to stored nuclear waste. Likewise infants and children bear a higher risk than do adults. According to recent US government estimates, a child is three to six times more likely than an adult to contract cancer when both have been exposed to one rad of total body radiation.[50]

Because of the health hazard borne by these high-risk groups as a likely result of nuclear waste storage, the failure to consider waste management costs increases the likelihood that the actual costs and benefits of nuclear fission generation of electricity will not be known. This omission also precludes providing a quantitative framework according to which those who now bear a disproportionate cost may be equitably compensated, and according to which the desirability of taking such a risk may be accurately assessed.

Because the financial burden of radioactive waste storage is excluded from economic analyses of nuclear power and simply taken on by the government, the taxpaying public is also the victim of an inequitable practice. The public is accepting the debts both of the nuclear industry and of the subset of persons who receive fission-generated power. Hence this practice is not only inconsistent with ethical policies prescribed by the US government, but also contrary to principles of equity. If the public as a whole bears the cost of waste storage, but only a subset of society receives the benefits of atomic power, then the costs and benefits of nuclear generation of electricity are not borne equitably. Allowing such a situation to continue means that policy regarding nuclear waste storage is implicitly founded on the presupposition that equity need not be served. This is a particularly dangerous presupposition, since US constitutional rights guarantee equal justice under law. As was pointed out in Chapter Two, equity is demanded both by the US Bill of Rights and by several rational considerations.[51]

2.2.2. *The Preclusion of Rational Choice*

Another ethical assumption, implicit in failing to account for radioactive waste, is that one is allowed to contract debts even when he has no assurance that he will be able to repay them. Such an assumption, contrary to most notions of fair play, has been made implicitly by the US Nuclear Regulatory Commission; its recent ruling implies "that it was not obligated to make a definitive finding that wastes generated by a reactor could be safely stored or disposed of as a prerequisite to allowing a reactor to operate".[52] Hence the Commission has presupposed that it is moral to allow industry to charge the debt of radioactive waste to the future, without certain knowledge that the debt can be paid by safe storage. This assumption really comes down to the belief that it is ethically permissible to subject society to a risk even when the degree of that risk is wholly unknown. US policymakers have stated explicitly that "the level of risk that the population might be subjected to can only be determined by assessments performed at actual repository sites". This information cannot be obtained ahead of time owing to "geological heterogeneities which preclude the transfer of basic earth science information . . . from one site to anther".[53]

Since the risk arising from nuclear waste storage is unknown in large part, the morality of permitting this storage is also questionable. Obviously, the morality of imposing a certain risk on the public is contingent both on how equitably the hazard is divided, on how well understood the risk situation is, on whether the hazard is accepted voluntarily, and on whether the consequences of the risk are common ones.[54] In imposing nuclear waste on the public, US policymakers are clearly assuming that their action is ethical, despite the facts that the risk is not borne equitably, the risk situation is not well understood, the danger has not been voluntarily accepted by the public, and the consequences of the risk are among the most dangerous known to man.

What makes such a policy questionable, from an ethical point of view, is not that only voluntary risks ought to be adopted, or that only fully understood risks ought to be accepted, or even that only equitably distributed risks ought to be taken. Clearly, the voluntariness, equity of distribution, and knowledge of a risk are parameters which have to be assessed in the light of the benefits of nuclear technology. Obviously, there are conceivable situations, within which principles of equity might be justifiably violated (e.g., if a completely equitable distribution of costs and benefits were impossible). The real problem is that current analyses of nuclear power do not provide a framework for knowing whether the waste storage situation is such a case. Likewise, there are probably situations within which unknown risks may be justifiably imposed on the public (e.g., if it is clear that not imposing them will very likely lead to even worse consequences). Whether this also is the case with waste storage is not known, because of the incompleteness of the assessments of nuclear energy.

The very fact of our ignorance regarding the many costs of radioactive waste storage suggests that we ought to take a morally conservative position regarding it. When faced with a choice between, for example, the (possibly) permanent hazards caused by nuclear wastes, and the (known) short-term risks imposed by pollution from the coal fuel cycle, one ought to take the safer course of the two. The morally conservative course, dictated both by equity and by current US public policy (The National Environmental Policy Act of 1969), is built on the assumption that members of future generations have the same rights as we do. If this is the case, then the more ethical choice (of the two examples provided) is to accept short-term risks so as to avoid long-term ones.

As it is, however, the ethical problem (with failure to include the social, medical, and financial costs of radioactive waste storage in analyses of nuclear power) is that such a policy precludes rational ethical analysis of some costs and benefits, and therefore rational ethical analysis of energy policy. It may or may not be desirable to generate nuclear waste. What is clearly undesirable, for both economic and ethical reasons, is to eliminate the possibility of assessing accurately whether such generation is either economical or ethical.

2.2.3. *The Acceleration of Social Costs*

As already suggested, perhaps one of the gravest ethical consequences of public policy-making grounded on the argument from
ignorance is the likelihood that the social costs of nuclear waste
storage will increase. Those who produce radioactive wastes are
required neither to bear the financial burden of them nor to
include them in their economic analyses. Since the US government
officially owns all radioactive wastes from commercial reactors,
individual utilities are neither required to bear the cost of long-
term storage nor liable for any health effects caused by exposure
to the wastes.[55] Within a total current liability limit of $560
million, the US government and private insurers provide insurance
to nuclear industries for coverage of all accidents occurring prior
to the beginning of the storage phase of radioactive wastes, at
which time they become the property of the government.[56]

If the social and economic costs of waste storage are not
computed prior to assessing the costs and benefits of nuclear
technology, then it is highly possible that such wastes will not be
effectively regulated and safeguarded. In fact, as was mentioned
previously, nuclear wastes have been produced and stored for
almost forty years in the US without criteria for determining
acceptable levels of exposure from all the radioactive isotopes
involved.[57] Waste is currently stored **throughout** the US, for
example, even though there are no criteria for determining how
much radiation can legally escape. Federal regulations simply
require that radiation be kept "as low as is practicable". There
is also no price estimated, per acre of land required for **permanent**
radioactive storage, and no price determined, on a yearly basis, for
perpetual control of nuclear waste disposal sites.[58] Although the
risk from storage facilities has been termed 'negligible' by the US
government, and although all storage sites have been licensed as
'safe', the National Academy of Sciences has stated unequivocally

that "none of the operating [nuclear waste storage] sites is geolog-
ically suitable for storage of radioactive waste in perpetuity".[59] US
Geological Survey scientists have pointed out that, even at the time
that waste disposal sites were opened, their lack of suitability (e.g.,
because of fissures in the rock and the likelihood of ground water
contamination) for receiving radioactive wastes was recognized,
but that siting decisions were made largely on the basis of "the
economics of handling the material".[60]

The US Environmental Protection Agency has charged that waste
management facilities in the US need to be upgraded, but that this
will be impossible until a "comparison of benefits and costs of
alternative [waste storage] programs" is presented. Hence, in the
absence of detailed comparative analyses of waste management
programs, the tendency has been not to regulate waste manage-
ment/storage practices as strictly as the EPA believes should be
done.[61] For example, it is common practice in the US to release a
number of long-lived radioactive isotopes directly to the soil in
the form of liquid; in fact significant amounts of plutonium have
been 'disposed of' in this way.[62] This practice has already resulted
in contamination of the ground water and will unquestionably
cause a greater incidence of cancer and genetic damage among
remote populations who might eventually be exposed to contaminated ground water.[63] But since ground water contamination
(although essentially permanent) takes place over a long period of
time, and since economic analyses of nuclear power do not include
calculations of the long-term effects of the cumulative radioactive
contamination of the environment caused by waste storage/man-
agement,[64] there is no coherent justification for prohibition of
such practices.[65] Hence the absence of detailed cost-benefit analyses of nuclear waste management programs is likely to result in
the public's bearing higher social costs in the form of greater exposures to radiation. But if the magnitude of these social costs
increases, then it can only mean that the public is involuntarily
bearing an even greater, and therefore possibly less equitable,

burden in the risk-benefit tradeoff of nuclear energy generation. Without adequate economic data on waste storage costs, it is impossible to know whether these social costs, including inequity, are really worth it. It is also impossible to engage in an accurate ethical analysis of alternative energy policies.

3. Conclusion

The ethical and economic problems of imposing nuclear waste storage on the public, without first doing a thorough study of the costs and benefits to be expected, are well-illustrated by an example now classic among ecologists. Recently the Aswan Dam was built in Egypt. The purpose of the dam was to generate badly needed electricity and to provide water for irrigation of the lower Nile basin. Before the dam was begun, however, there was no thorough study of the costs and benefits to be expected, since Egyptian policymakers were convinced that the benefits of electricity and irrigation were so great that they would outweigh whatever costs ensued. Ignoring the ecological principles that everything is interconnected and that there are no ecological and economic gains obtained without sustaining some losses, they built the dam. Unforeseen costs of the project have put the benefits of the dam in question. The former flood plains are no longer fertilized by silt from the Nile, and artificial fertilizers how have to be used. The controlled irrigation has salinated the soil. The sardine catch in the nearby Mediterranean has decreased by 97% because the sea no longer receives flood-borne nutrients, and schistosomiasis (a debilitating disease dependent on snails) has reached epidemic proportions because of the constant supply of water (and therefore snails) provided by the irrigation.[66]

Just as the monetary and ethical costs of a new fertilizer industry, an expanded medical program, a soil reclamation service, a decreased fish catch, and an increase of disease were ignored by the Egyptians, so also the monetary and ethical costs of waste storage have been largely ignored in US public policy assessments of nuclear energy. Both these omissions have resulted in serious economic, ethical, and ecological consequences. Such failings cannot unequivocally be said to be either unethical or methodologically unsound; in fact, they are understandable in the light of our pursuit of abundant energy and the impossibility of ever obtaining complete knowledge about any situation. Nevertheless, public policy regarding both issues clearly warrants further study. In the case of nuclear fission, ignoring a key cost of the technology clearly precludes accepting the conclusions recently drawn by US policy-makers, viz., that nuclear power is cheaper and less environmentally damaging than some alternative forms of electrical energy, and that the atomic energy/equity tradeoff is clearly justified. There are several reasons to believe not only that these conclusions might be epistemologically and ethically unwarranted, but also that they are ecologically unsound. The solution, however difficult, can only be either to avoid drawing conclusions about the relative benefits and costs of nuclear power or to internalize the cost of waste management and therefore to 'ecologize' technology assessments. Either alternative would lead to a more desirable framework for making public policy regarding nuclear power.

Notes

[1] Quoted in US Energy Research and Development Agency, Draft Environmental Statement, *Waste Management Operations: Savannah River Plant, Aiken, South Carolina* (ERDA-1537), US Government Printing Office, Washington, D.C., October, 1976, p. K-62; hereafter cited as:'ERDA-1537-DRAFT'.
[2] This is admitted clearly in US government publications. See, for example, US Environmental Protection Agency, 'Criteria for Radioactive Wastes', *Federal Register* **43**(221), (November 1978), 53263; hereafter cited as: 'Wastes'.
[3] J.M. Deutch and the Interagency Review Group on Nuclear Waste Management, *Report to the President* (T1D-2817), National Technical Information Service, Springfield, Virginia, October 1978, p. iii; hereafter cited as: IRG, *Report*.
[4] IRG, *Report*, p. xxiii. For a discussion of 'high-level waste', see Section 1 of this chapter.
[5] US EPA, 'Wastes', 53263.

[6] G. Hart, 'Address to the Forum', US Environmental Protection Agency, *Proceedings of a Public Forum on Environmental Protection Criteria for Radioactive Wastes* (ORP/CSD-78-2), US Government Printing Office, Washington, D.C., May 1978, p. 5; hereafter cited as: 'ORP/CSD-78-2'.

[7] IRG, *Report*, p. iii.

[8] B. Kiernan *et al.*, Legislative Research Commission, *Report of the Special Advisory Committee on Nuclear Waste Disposal*, No. 142, Legislative Research Commission, Frankfort, Kentucky, October, 1977, p. 3; hereafter cited as: 'LRC'.

[9] Kiernan *et al.*, 'LRC', p. 3.

[10] US Environmental Protection Agency, *Considerations of Environmental Protection Criteria for Radioactive Waste*, US EPA, Washington, D.C., February 1978, p. 1; hereafter cited as: *Criteria* (also cited in note 9, Chapter 2).

[11] US EPA, *Criteria*, p. 9.

[12] IRG, *Report*, iv; see also US EPA, *Criteria*, p. 23, where it is explained that many of the radioactive wastes will have to be isolated 'permanently' (for one million years) from the biosphere. See also *Criteria* (note 10 above). For more information regarding the necessity of permanent containment, see US Energy Research and Development Administration, *Final Environmental Statement: Waste Management Operations, Hanford Reservation, Richland, Washington*, Vol. 1 (ERDA-1538), National Technical Information Service, Springfield, Virginia, October 1975, p. X-179; hereafter cited as: 'ERDA-1538'.

[13] 'ERDA-1538', p. X-159.

[14] US AEC, *Comparative Risk-Cost-Benefit Study of Alternative Sources of Electrical Energy* (WASH-1224), US Government Printing Office, Washington, D.C., December 1974, p. 4-30; hereafter cited as 'WASH-1224'.

[15] 'WASH-1224', p. 4-14. According to the *Code of Federal Regulations* 10, Part 20, US Government Printing Office, Washington, D.C., 1978, p. 184 (hereafter cited as: '10 CFR 20'), one 'rad' of radiation is "a measure of the dose of any ionizing radiation to body tissues in terms of the energy absorbed per unit mass of the tissue".

[16] W.D. Rowe, Deputy Assistant Administrator for Radiation Programs, Environmental Protection Agency, in testimony before the Committee on Commerce, Science, and Transportation, *Radiation Health and Safety*, US Senate, 95th Congress, First Session, June 16, 17, 27–29, 1977 (No. 95–49), US Government Printing Office, Washington, D.C., 1977, pp. 102–103; hereafter cited as: 'Committee on CST'. See also 'WASH-1224', p. 4-2, where US government researchers state that "it is impossible to state *a priori*" whether a particular case of cancer or genetic damage was caused by medical, natural, or nuclear-fuel-cycle radiation.

[17] Kiernan *et al.*, 'LRC', pp. ix–17, and US ERDA, 'ERDA-1538', Vol. 2, pp. 11.1–H–1 through 11.1–H–4.

[18] Hart in US EPA, 'ORP/CSD–78–2', p. 6.
[19] US ERDA, 'ERDA-1538', Vol. 1, p. X–74. See Chapter 2, notes 13 and 16.
[20] US ERDA, 'ERDA-1538', Vol. 1, p. II. 1–57.
[21] US ERDA, 'ERDA-1538', Vol. 1, p. III. 2–2.
[22] US ERDA, 'ERDA-1537', *Waste Management Operations: Savannah River Plant, Aiken, South Carolina* (UC-2-11-70), National Technical Information Service, Springfield, Virginia, September, 1977, pp. III–20 and IV–2; hereafter cited as: 'ERDA-1537'. See also US ERDA, 'ERDA-1536', *Waste Management Operations, Idaho National Engineering Laboratory, Idaho*, National Technical Information Service, Springfield, Virginia, September, 1977, p. E–41; hereafter cited as: 'ERDA-1536'.
[23] IRG, *Report*, pp. iv, 34. US ERDA, 'ERDA-1538', Vol. 1, p. X–203, also confirms the fact that waste storage facilities will not be available at least until 1995.
[24] For periods "greater than one to two hundred years, uncertainties dominate predictive ability". US EPA, *Criteria*, p. 26.
[25] Dr. David Hall, cited in the Hearings Before the Subcommittee on Oversight and Investigations of the Committee on Interstate and Foreign Commerce, House of Representatives, *Nuclear Waste Management Disposal*, 95th Congress, First Session, July 29 and August 1, 1977 (No. 95-67), US Government Printing Office, Washington, D.C., 1977, p. 53; hereafter cited as: 'Committee on IFC'.
[26] This latter policy is under consideration as part of a US attempt to curb nuclear proliferation of weapons. For more information on this and on US ownership of wastes generated within the US, see Committee on IFC, pp. 2, 33; and IRG, *Report*, p. 55.
[27] AEC, 'WASH-1224', pp. 1-22 through 1-24; see also pp. 5-29 through 5-31, 5-43, and 5-48 through 5-50.
[28] 'Committee on IFC', p. 38.
[29] US ERDA, 'ERDA-1537', p. 1-15; US ERDA, 'ERDA-1538', Vol. 1, pp. X–72, I–5 and I–25; 'Committee on IFC', p. 140.
[30] AEC, 'WASH-1224', pp. 1-23, 1-24, and 5-5.
[31] IRG, *Report*, p. 68.
[32] IRG, *Report*, p. 4.
[33] See notes 17 to 22.
[34] See note 19.
[35] These calculations are based on government data cited in note 15 and found in AEC, 'WASH-1224', p. 3-83. They are based on the premise that one rem of radiation causes 1/5000 cancers and 1/500 genetic deaths.
[36] AEC, 'WASH-1224', p. 5-5.
[37] See note 32.
[38] This is a plausible assumption since approximately one half of all high-

level wastes now stored in the US are from defense projects, whereas the other half is from reprocessing spent commercial reactor fuel (IRG, *Report*, p. D–19). Likewise, commercial transuranic and low-level wastes from reactors now amount to approximately 25% of all transuranic and low-level wastes in the US, but by the year 2000, commercial transuranic and low-level wastes from reactors will constitute 75% of all low-level and transuranic waste in the US. (IRG, *Report*, pp. D-11; D-12; D-14.)

[39] Testimony of L. V. Gossick, Executive Director for Operations, Nuclear Regulatory Commission, indicates that there are approximately 65 commercial reactors now in operation and approximately 70 now under construction. 'Committee on IFC', p. 103.

[40] IRG, *Report*, p. 36. See also US ERDA, 'ERDA-1538'. US government officials have clearly indicated that "it is doubtful if any nuclear facility could be built which did not release at least some radioactivity". (Kiernan *et al.*, 'LRC', p. 25; EPA, *Criteria*, pp. 41, 23.)

[41] See note 30.

[42] IRG, *Report*, p. xx. Numerous other US government documents mention the lack of criteria for 'acceptable' levels of radioactivity, despite the fact that all radiation is carcinogenic and mutagenic. Some of the more important areas in which contamination from storage sites has been ignored include food-chain effects, dose factors, long-term dose commitment, and terrestrial and aquatic ecosystem involvement. (See US ERDA, 'ERDA-1537-DRAFT', pp. K–12, K–4; US ERDA, 'ERDA-1538', Vol. 1, p. VI-1; and Kiernan *et al.*, 'LRC', p. 21.)

[43] J. J. Berger, *Nuclear Power*, Ramparts, Palo Alto, 1976, p. 112; hereafter cited as: *Nuclear Power*.

[44] Berger, *Nuclear Power*, p. 150.

[45] US ERDA, 'ERDA-1538', Vol. 1; see especially p. X–28.

[46] A.Z. Roisman, Natural Resources Defense Council, in testimony before the 'Committee on IFC', p. 53. See also note 18.

[47] IRG, *Report*, p. vi.

[48] Several authors have provided excellent rational justifications for our obligations to members of future generations. See R.M. Green, 'Intergenerational Distributive Justice and Environmental Responsibility', *Bioscience* 27 (4), (April 1977), 260–65; and Daniel Callahan, 'What Obligations Do We Have to Future Generations?' *The American Ecclesiastical Review* 164(4), (April 1971), 265–80.

[49] National Environmental Policy Act, cited in EPA, 'Criteria for Radioactive Wastes', *Federal Register* 43(221), Part 9 (November 1978), 53267–62.

[50] AEC, 'WASH-1224', p. 4-14.

[51] There are numerous philosophical frameworks within which this presupposition (that equity need not be served) may be questioned. Some of the best contemporary arguments for the principle of equity may be found in John Rawls, *A Theory of Justice*, Harvard University Press, Cambridge, 1971,

pp. 100–114, 342–50. See Sections 3.1. and 3.2. of Chapter Two herein.

[52] 'Committee on IFC', p. 52.

[53] IRG, *Report*, p. 28.

[54] This point is also made by W. W. Lowrance, *Of Acceptable Risk: Science and the Determination of Safety*, William Kaufmann, Los Altos, 1976, p. 36.

[55] 'Committee on IFC', pp. 2, 33; and IRG, *Report*, p. 55.

[56] R. Wilson, 'Nuclear Liability and the Price-Anderson Act', *Forum* XII(2), (Winter 1977), 612–21.

[57] See note 42.

[58] US ERDA, 'ERDA-1538', p. X–187.

[59] US ERDA, 'ERDA-1538', pp. II. 3–78 and X–188.

[60] Kiernan *et al.*, 'LRC', p. 15; see also US ERDA, 'ERDA-1537-DRAFT', p. II–120, p. K–60, p. K–54 through K–56, K–69; US ERDA, 'ERDA-1536', pp. X–168, X–170 and X–181; US ERDA, 'ERDA-1538', Vol. 1, pp. II. 3–77; II.1–44–45; and X–195; F.E. Stubblefield and E.B. Jackson, Atomic Energy Commission, *Improved Control of Radioactive Waste at Hanford* (WASH-1315), Government Documents, Washington, D.C., June 1974, pp. 4-7 and 19; hereafter cited as: 'WASH-1315'.

[61] US ERDA, 'ERDA-1538', p. X–71.

[62] Stubblefield and Jackson, 'WASH-1315', pp. 5–7, 19; US ERDA, 'ERDA-1538', Vol. 1, p. II.3–77.

[63] US ERDA, 'ERDA-1538', Vol. 1, pp. II.3–77 and II.1–44ff.; US ERDA, 'ERDA-1536', p. X–168 through X–170; US ERDA, 'ERDA-1537-DRAFT', p. K–69.

[64] See note 17.

[65] US ERDA, 'ERDA-1538', p. X–163. See notes 17, 27, 42, 45, and 53.

[66] This example was given by Garrett Hardin, 'To Trouble a Star: The Cost of Intervention in Nature', in *Environment and Society* (ed. by R. T. Roelofs, J. N. Crowley, and D. L. Hardesty), Prentice-Hall, Englewood Cliffs, N.J., 1974, pp. 120–21.

Chapter Four

Core Melt Catastrophe and Due Process

Despite more than three decades of government-subsidized promotion and regulation of nuclear fission technology, a number of questions remain problematic. Some of these are economic, a few are purely scientific, while others are largely political. Nearly all of them raise troubling ethical considerations. From a philosophical point of view, one of the most interesting of these questions is whether the US government ought to continue to protect the nuclear industry by imposing a limit on liability in the event of a catastrophic reactor accident. Answering this query demands, not only that one analyze the ethical presuppositions underlying assessment and regulation of nuclear technology, but also that, in general, one evaluate government guardianship of the utility industry. This latter task was urged long ago by Juvenal, when he pondered: *"Quis custodiet ipsos custodes?"* ("Who shall guard the guardians themselves?").[1]

Society has always feared governments of bureau administrators; their power typically has led to corruption and to rule by men instead of laws.[2] Federal regulation of nuclear energy presents a particularly difficult and important subject for analysis, both because the complexity of the technology virtually necessitates control by a regulatory elite rather than directly by the body politic, and because abuses within government and industry have potential for widespread destruction. In this chapter I will investigate the philosophical presuppositions implicit in the government guardianship of nuclear technology. Specifically I will evaluate the logical, methodological, and ethical theories underlying the Price-Anderson Act. All too often the philosophical theories underlying such decision-making have gone uncriticized because their proponents claimed their views were not theory-laden, but merely common-sensical or 'just business'.[3] Hence the philosophy implicit in technological guardianship has sometimes enjoyed an undeserved power. As a consequence, fundamental principles of freedom, justice, and security often have been compromised. As Albert Schweitzer put it:

So little did philosophy philosophize about civilization that . . . in the hour of peril the watchman who ought to have kept us awake was himself asleep and the result was that we put up no fight at all on behalf of our civilization.[4]

Although the debate over nuclear energy is a many-sided one, the basic issue is the adequacy of federal guardianship, since the government has assumed almost absolute authority for making the tradeoffs between energy demands and the radiation risks of nuclear power. If the thesis of this chapter is correct, however, there are strong philosophical reasons both for the people, rather than the federal government, to play the role of the Solomon who must weigh the goods and evils arising from nuclear power, and for the balance to shift in the character of these public policy assessments.

1.   The Price-Anderson Act

To understand the significance of the Price-Anderson Act, one must know something of the history of nuclear energy. As was pointed out in Chapter One, the nuclear industry dates to the forties, when the US spent five years and $ 2 billion to build the first atomic bombs during World War II. Thereafter the government took twenty years and more than $ 100 billion in subsidies to develop the first power reactors used to generate electricity. The reasons for beginning to develop nuclear fission in the forties and fifties were that the military wanted bombs, and the government was eager to find uses for the peaceful atom. Both the Cold War and 'Atoms for Peace' thus provided a rationale for continuing the expansion of nuclear technology and hence for obtaining the weapons-grade plutonium as a reactor by-product.[5]

As was also explained in the first chapter, one legacy of the arms race is that US nuclear technology is built around an enriched-uranium reactor much more susceptible to catastrophic accident than a natural-uranium one. A second consequence of the military origins of the current nuclear industry is that the US has sponsored a major technology without ever assessing whether it is a desirable means to a quite different end, viz., electrical power rather than nuclear warheads.

Industry, however, did evaluate the desirability of using atomic power to generate electricity. As was pointed out in Chapter One, power companies refused initially to invest their funds in such a dangerous technology, since the legal consequences of a reactor accident could destroy the assets of a utility many times over.[6] Because private firms would not provide more than a small fraction of insurance coverage, the government was forced to protect the industry itself. In 1954, Congress passed the Atomic Energy Act, whose object was "to promote the private development of atomic energy".[7] Three years later the Price-Anderson Act, as Section 170 of the Atomic Energy Act, became law. Under its provisions, (1) the nuclear power plant owner was required to obtain $ 60 million of insurance, viz., all that private insurers would allow; (2) the US government provided an additional $ 500 million indemnity protection for each nuclear 'incident'; and (3) liability of all persons, in the event of an 'incident', was limited to $ 560 million; the act specifically insured that the law will "hold harmless the licensee and other persons indemnified . . . from public liability arising from nuclear incidents in excess of the level of financial protection required of the licensee".[8]

The unwillingness of private insurers to provide sufficient coverage, together with the nuclear industry's continuing insistence that a government-guaranteed liability limit is a necessary condition for their remaining in the business of nuclear generation of electricity, have had an unforeseen consequence.[9] Congress has been forced to extend the Price-Anderson Act, justified merely as a 'temporary measure', beyond its original ten-year period.[10] In 1965, the act was extended for another ten years.

As a result of amendments made in 1965, the act now provides that a waiver of defenses against all liable persons is required, provided that an 'extraordinary nuclear occurrence' has taken place. The Nuclear Regulatory Commission has the power to define such occasions, and defenses are waived for these incidents. The injured person is required only to show that an accident at a nuclear power plant caused his injury, but not that the owners/operators were negligent. Amendments also made in 1965 provide for the immediate partial compensation to the victims without their having signed a release.[11]

Even though mounting evidence indicated that the government was not protecting the public adequately against the threat of nuclear accidents,[12] the Price-Anderson Act was extended again in 1975 to cover the period up to August 1, 1987. At this time Congress also passed two major amendments to the act.

First, it provided a 'deferred premium plan' according to which, after several more decades have passed, there will be a gradual substitution of industry-financed insurance for government-financed coverage of nuclear liability. Currently private insurers are willing to offer $125 million in coverage to the nuclear industry, and the government provides the remaining $435 million in coverage, i.e., that up to a total of $560 million. The substitution, of industry-financed insurance for government-financed coverage, will be accomplished by a 'deferred premium' plan which will operate only when an 'extraordinary nuclear occurrence' takes place. In this event, each licensee will be assessed a prorated share of the damages above $ 125 million, provided no premium is less than $ 2 million or more than $ 5 million. The level of the premium is set, within these constraints, by the NRC. The date on which government subsidy is phased out depends on the level at which the premium is assessed and on the number of power plants operating. If all currently operating plants (65) were assessed at the maximum $ 5 million, the total maximum amount now available from the deferred premium plan would be $ 325 million. This means that the government would still have to supply a $ 110 million subsidy, since the liability limit is $ 560 million.[13]

A second amendment stipulates that, if at least twenty-six more power plants are built, and if the maximum ($ 5 million per licensee) premium is assessed, then the nuclear industry's liability might rise (at most) at a rate of $ 5 million for each plant built after the next twenty six. For example, if 26 more plants were built by 1987, then in that year, the liability limit could be no greater than $ 565 million. If 27 plants are built by 1987, then in that year the liability limit could not exceed $ 570 million. In any event, the ceiling of $ 560 million will rise only if at least 26 more plants are built. If the 70 plants currently under construction eventually begin operation, then the total liability limit could range from $ 560 million to $ 800 million (($ 5 million x 135 plants) + ($ 125 million)), provided that the government required all plant licensees to pay the maximum allowable premiums.[14]

2. Philosophical Difficulties in the Price-Anderson Act

From the preceding account of the Price-Anderson Act, it is easy to understand why this law has been called "the single most important legislative contribution to the development of the nuclear industry."[15] Moreover despite recent drastic cutbacks in orders of nuclear reactors, a significant percentage of this country's electric capacity is now provided by atomic plants, and others continue to be built. Hence it is imperative that the philosophical presuppositions of the Price-Anderson Act be analyzed.

2.1. LOGICAL PROBLEMS WITH PUBLIC POLICY GOVERNING LIABILITY

On a purely logical level, there is question whether the Price-Anderson Act is consistent, both with certain presuppositions underlying other government laws and with accepted factual and scientific assumptions. The most obvious sense, in which this is the case, has to do with the Brookhaven Report, WASH-740.

2.1.1. *Inconsistency between Damage Estimates and Liability Limits*

On the one hand, "the Price-Anderson Act was designed to meet two basic objectives: (1) to . . . satisfy public liability claims [of up to $560 million] in the event of a catastrophic nuclear accident and (2) to remove the deterrent to private industrial activity in atomic energy presented by the threat of enormous liability claims."[16] On the other hand, the government document, WASH-740, concludes that *property damage* alone, in the event of a nuclear accident, could go as high as $17 billion, apart from the 45,000 immediate deaths and the 100,000 later deaths, injuries, and cancers predicted.[17] Moreover since nuclear plants today are approximately five times the size of those whose accidents were described in the original Brookhaven Report, it is likely that consequences of a core melt would be even more severe than those calculated in WASH-740.

The Brookhaven Report was originally requested by the Joint Commission on Atomic Energy because the Commission wanted positive safety conclusions "to reassure the private insurance companies" so that they would provide coverage for the nuclear industry. Since even the conservative statistics of the report were alarming, it was suppressed and its data were not made public until almost 20 years later, when a suit was brought as a result of the Freedom of Information Act.[18] It seems likely that, were the Brookhaven conclusions released prior to passage of the Price-Anderson Act, this legislation might not have been accepted. Now that the data are known, they provide a firm basis for challenging the Price-Anderson Act.

Clearly either the Price-Anderson Act is to "satisfy public liability claims", as the Atomic Energy Commission says, or it is not. If it is to meet these claims, then its liability limits must be consistent with governmental damage estimates, and not merely 3% (i.e., $560 million divided by $17 billion) of them. If the Act is not to compensate for damages, then Price-Anderson need not be amended, but government officials must retract the claim that it will "satisfy public liability claims". (In a later section of this chapter I will argue why the most ethical decision is to raise the liability limits of Price-Anderson.)

A second disturbing fact is that the government estimates of the nuclear risk, presupposed in the Price-Anderson Act, are inconsistent with the estimates given by private insurers. Clearly either the $435 million government insurance subsidy (provided to the nuclear industry above the $125 million available through private insurance companies) represents a risk the government ought to take with the taxpayers' money, or it does not. If such a risk is economically desirable, then a profit can very likely be made on nuclear insurance and the taxpayers' money probably will not be lost. In this case, private industry ought to be ready to invest heavily in insuring the nuclear industry. However, facts do not bear this out. After more than twenty years during which nuclear reactors have generated power, the maximum insurance obtainable for them through private insurance pools is $125 million. This is especially interesting, since private insurance groups have provided, for example, up to $ 500 million in coverage for Pan American Airlines.[19] If nuclear plants are safe, and if private insurers are *incorrect* in failing to provide adequate coverage for them, then why is it reasonable for the government to limit liability for a reactor accident? On the other hand, if nuclear plants are not safe, and if private insurance companies are *correct* in their unwillingness to take such a risk,[20] then clearly citizens ought not to be limited in their right to collect damages on their property.

Several facts will indicate precisely why the assumptions of the insurance industry are incompatible with the actuarial presuppositions underlying the Price-Anderson Act. The premiums currently paid to private insurance pools for $ 125 million worth of coverage are based on the assumption that the chance of a serious accident is about 1 in 20.[21] Price-Anderson (government) statistics, on the other hand, are based on the assumption that the probability of a serious accident is about 1 in 17,000.[22] Likewise the 'deferred premium', established through the Price-Anderson Act, has been set by the government at a maximum of $ 5 million per reactor, whereas private insurers maintain that, to cover all damages resulting from a serious nuclear accident, an annual premium of at least $ 23.5 million per plant per year would have to be assessed.[23] Hence, at best, the actuarial presuppositions underlying the Price-Anderson liability limitation are inconsistent with fairly reliable insurance data. In a later section of this chapter I will argue why, for methodological reasons, the statistics from private insurers are probably more reliable than the government figures.

Another apparent inconsistency arises, if one examines the reasons usually employed to justify the Price-Anderson Act. According to its proponents, the liability limitation will never actually deprive a person of the right to sue for property damages, because a serious nuclear accident will never occur. However industry refused to enter the nuclear field until a liability limit was established through the Price-Anderson Act.[24] Now either nuclear power is safe and catastrophic accidents are impossible, in which case no limit on liability is needed to protect the nuclear industry from bankruptcy; or, on the other hand, nuclear power is not safe and catastrophic accidents are possible, in which case a limit on liability is needed to protect the nuclear industry from bankruptcy. If the limitation is needed, it can only be so because successful claims can be made against the industry. But successful claims can be made against the industry only when injury can be shown to be the result of a nuclear accident. And if this can be shown, nuclear power is not safe. Hence one cannot argue consistently, both that there is a need for a limit on nuclear liability and that nuclear reactors are safe.[25] Perhaps this logical difficulty is the reason why some proponents of nuclear energy have argued that the industry needs billions, not millions, of dollars of coverage.[26]

2.1.2. *Incompatibility with the Energy Reorganization Act*

A final respect in which the Price-Anderson Act may be said to be questionable on logical grounds is in its relationship to the Energy Reorganization Act of 1974. In this legislation, Congress declared that "all possible sources of energy be developed".[27] Yet, in the very next year, Congress extended the Price-Anderson Act and reaffirmed its subsidy of nuclear power. As several Congressmen have noted, this continuation of the act served the "effect of concentrating Federal funds on one energy option, namely nuclear".[28]

Almost two hundred billion dollars already have been invested in nuclear power, through subsidies of the Price-Anderson Act, reprocessing, waste storage, research and development, enrichment, and site restoration.[29] 83% of federal energy monies have been spent for nuclear fission research, and all the other energy options have shared the remaining 17% of funds.[30] Although there is no formal inconsistency between the Energy Reorganization Act and the Price-Andereson Act, it does not seem logical both to continue massive subsidies of the most heavily-supported energy alternative and at the same time, to urge development of all energy options.

2.2. METHODOLOGICAL PROBLEMS WITH ASSESSMENTS OF SAFETY

In addition to the purely logical questions that can be raised about the presuppositions of the Price-Anderson Act, relative to other theses supported by government, industry, and private insurers, this legislation also raises a number of methodological issues. Perhaps the most obvious premise, upon which the liability limitation is based, is that the likelihood of a catastrophic nuclear accident is so remote as to be almost impossible.[31] The government's reasoning here is that since a serious accident is unlikely, massive insurance protection is not needed for the public; therefore it is reasonable for the Price-Anderson Act to establish a liability limitation.[32] This thesis, (A), that the probability of a catastrophic reactor accident is so unlikely as to be impossible, depends upon a number of other propositions which are debatable on methodological grounds. Some of the more important of these are the following: (A_1), The calculation of accident probabilities in the government document WASH-1400, known as the Rasmussen Report, is based upon reasonable mathematical assumptions. (A_2), No data relevant to computation of these accident probabilities was suppressed. (A_3), The emergency core cooling system is reliable. (A_4), Even if a core melt did occur, most people could be evacuated from the area surrounding a nuclear power plant. In the following paragraphs I will outline my reasons for arguing why the scientific methodology upon which (A_1) ... (A_4) is based is doubtful, and consequently why thesis (A) and, therefore, the liability limitations of the Price-Anderson Act, are open to question.

2.2.1. *Mathematical Assumptions Underlying Accident Probabilities*

Thesis (A_1) has been used, perhaps more than any other premise, to support public confidence, both in this nation's nuclear fission technology and in the liability provisions of the Price-Anderson Act.[33] There are a number of facts, however, that suggest that (A_1) is not the case. For one thing, both the EPA and the American Physical Society, the most prestigious group of physical scientists in the world, conducted reviews of WASH-1400, the Rasmussen Report, and came to the conclusion that "the calculational methods used in both the research program and the regulatory function [are] fairly unsatisfactory . . . . The consequences of an accident involving major radioactive releases have been underestimated."[34]

The general problem with the mathematical methodology used in WASH-1400 is that its main technique is 'fault tree analysis', developed many years ago by the Department of Defense and NASA, both of whom consider it erroneous, "obsolete", and merely "educated guessing".[35] Its difficulties lie in the fact that the method provides no way to account for all parameters contributing to an accident. For example, the probabilities of a serious accident were calculated without taking account of the aging and deterioration of the plant, especially the reactor vessel, and the effects of terrorism and sabotage. The APS also criticized, for example, the assumption "that persons downwind of a reactor accident would receive radiation only through the first day following the accident", and noted that correction of this error alone would increase the number of cancers and genetic damages resulting from an accident by a factor of 25. This deficiency in mathematical methodology, however, is just one of many. The APS noted that WASH-1400 has hundreds of such errors, e.g., the neglect of damages resulting from resource contamination through land and water, with consequent effects on 5 or 6 generations into the future; the neglect of effects of "the selective and much larger irradiation of special tissues, notably the lungs and thyroid". For these and other reasons, the damages predicted by the APS represent an increase over the Rasmussen statistics by a factor sometimes as great as 50. For example, WASH-1400 said there would be 310 fatal cancers, apart from immediate deaths and genetic damage, as a result of a reactor accident. The same prediction made by the APS was for between 10,000 and 15,000 fatal cancers.[36] Moreover, as other groups of scientists have pointed out, the Rasmussen figures are too low because they exclude delayed deaths and injuries, such as cancer and genetic damage appearing in later years or in subsequent generations; this omission is significant because delayed nuclear fatalities are thought to exceed prompt ones by a figure of 100 or more.[37]

Even where the mathematical methodology of the Rasmussen Report is not erroneous, it is misleading because of the incompleteness of the analysis. For example, the report estimates that the per-year, per-reactor probability of a core melt accident is one in 17,000 (see note 22). Even if it is correct, this figure is not as meaningful as the probability that, in the next thirty years, there will be a core melt accident in one of the plants now operating or currently under construction. This probability is one in four.

Using the Rasmussen probability (per year, per reactor) for a core melt, one can compute the probability that a core melt will occur in one of the 75 plants now operating during their 30-year lifetime. Employing the formula for the probability of mathematically independent events, P (a core melt in at least one of 75 reactors over a 30-year lifetime) = $1 - P$ (no core melt in any of 75 reactors over a 30-year lifetime), one obtains P (core melt) = $1 - (1 - (1/17,000))^{2250} = 1 - 0.8761 = 0.124$. Thus even if reactors now under construction and now planned are not built, there is still a 12% probability, or a 1 in 8 chance, of a core melt in the 30-year lifetime of one of the reactors now operating. Using the same formula and the Rasmussen probability, one can also compute the probability of a core melt in the 30-year lifetime of 151 plants (assuming 75 now operating and 76 under construction). This $P = 0.239$ or 24%. Thus for 151 reactors, the chance of a core melt is approximately 1 in 4.

In the light of the mathematical analysis just given, it is doubtful whether it is the case that (A_1), the calculation of accident probabilities in the Rasmussen Report is based upon reasonable mathematical assumptions. But if it is unlikely that (A_1) is the case, it is also questionable that (A), the likelihood of a catastrophic nuclear accident is so remote as to be impossible. (A) also appears unlikely on the basis of the 1979 accident at TMI (Three Mile Island), Pennsylvania. Although this crisis, itself, says nothing about probabilities, Norman Rasmussen, author of the famous Rasmussen Report, revealed some startling information about TMI. After the accident, he said that he had calculated the TMI event to have a probability of from 1 in 250 to 1 in 25,000 per reactor-year.[38] This does not sound disturbing until one does some calculation to show that, for the 151 reactors now operating or under construction, the 1 in 250 figure yields the conclusion that a TMI-type accident can be expected every other year in the US. Rasmussen, however, did not reveal this conclusion, although it can be obtained easily by employing the formula for the probability of mathematically independent events; the annual probability that another TMI will occur somewhere in the US is 0.46 or 46%.[39] Therefore if an accident cannot be said to be highly improbable, there are economic, political, and ethical reasons to question the desirability of the Price-Anderson liability limitation.[40]

2.2.2. *Suppression of Data Regarding Nuclear Hazards*

Of course if data relevant to the Rasmussen calculations were suppressed, or if WASH-1400 was not developed in an objective and disinterested manner, then it would be easy to understand the discrepancies between statistical conclusions of government and the scientific community. Although it is not possible to determine with certainty whether (A_2) is the case, i.e., whether no reliable data relevant to computation of these accident probabilities were suppressed, a number of facts shed light on this issue. First, as was documented in Chapter One, the Energy Reorganization Act of 1975 was adopted specifically to counteract the numerous problems arising because of the pro-nuclear industry bias of the Atomic Energy Commission. The Rasmussen Report was commissioned by the AEC and finished in 1974, just at the time that abuses in the AEC were beginning to come to light. Secondly, the government suppressed the results of the Brookhaven Report (WASH-740, completed in 1957 and updated in 1965) for almost twenty years until it was forced to release them under the Freedom of Information Act.[41] Both of these facts, along with suppression of data related to nuclear weapons' testing in the fifties,[42] indicate that there is strong reason to doubt the truth of (A_2).

Other considerations also suggest the likelihood that pro-industry bias has resulted in suppression of important material. For example, the Atomic Energy Commission, later the Nuclear Regulatory Commission, has given repeated permission, not only for site preparation and construction activities prior to any public hearing or construction permit for a nuclear plant,[43] but also for required environmental impact statements to be ignored.[44] Government regulators allowed a fast breeder reactor to be constructed, even though its Advisory Committee on Reactor Safeguards opposed the permit, and then "concealed that fact from Congress and the public".[45] They are well known for the facts that they will finance only speakers who are proponents of nuclear fission,[46] that they have 'punished' employees who have taken the safety aspects of their job too seriously,[47] and that they have censored government research reports critical of nuclear power.[48]

For all these reasons, there is at least a strong possibility to doubt (A_2) that no reliable data relevant to computation of accident probabilities were suppressed, and hence to question thesis (A), that the probability of a catastrophic reactor accident is virtually impossible.

2.2.3. *The Reliability of the Emergency Core Cooling System*

Another reason to question (A) is that statistical assumptions (about the low probability of a serious reactor accident) are based on an empirical thesis (A_3) whose validity has never been tested. (A_3), that the emergency core cooling system is reliable, is highly doubtful on grounds of scientific methodology; no full scale, empirical test of this cooling system has ever taken place.[49] Moreover all of the small, scale-model tests (six out of six) have failed.[50] Instead Nuclear Regulatory Commission officials have sanctioned computerized calculations of the probability that the system will work. All problematic possibilities obviously cannot be accounted for in a small physical model, and even fewer risk factors can be treated in a purely mathematical model.

The mathematical model has been used, not because scientists are convinced of its merits, but rather because the full-scale tests would be 'possibly very destructive' and would involve a great 'risk'.[51] The logical weakness of such a reason for failing to test the emergency core cooling system is obvious. If the system is effective, then a test of it would not be catastrophic. On the other hand, the only reason why a test would be destructive is if the system is not effective, which is *prima facie* evidence that it ought to be tested. As the recent loss-of-coolant accident (LOCA) at Three Mile Island (TMI), Pennsylvania reveals, a core melt may not have a low probability. Moreover, simply testing the emergency core cooling system for mechanical failures does not address one of the key contributors to the TMI accident: *human error*. As the President's report on TMI indicates, the operators failed to recognize the significance of mechanical indicators of the accident, and their supervisors in turn failed to recognize core damage.[52] This case illustrates that, even if the mechanical systems work perfectly, human error can cause a core melt. Hence there is strong reason to doubt thesis (A_3) and therefore strong reason to question whether a core melt has a low probability.

2.2.4. *Assumptions of Near-Complete Evacuation*

Even if a core melt did occur, however, one of the crucial presuppositions undergirding (A), the thesis of the low probability of a disastrous accident, is (A_4), that 90% of the people could be evacuated from the area surrounding the nuclear power plant. (A_4) is doubtful on methodological grounds for a number of reasons. For one thing, some power plants are located in extremely populous areas (e.g., the Indian Point reactors are 26 miles from the heart of Manhattan). Secondly, government scientists assumed that 90% evacuation would occur, despite their admissions that the range of evacuation would have to extend at least to 400 square miles in the case of a serious accident and that the maximum warning time prior to a major nuclear catastrophe would be only 2½ hours.[53] Thirdly, almost no evacuation plans have been evaluated and there is no solid, empirical data to support them. Where the plans have actually been tested, almost all have failed, as in the 'emergency response test' of the reactor in Toms River, New Jersey recently.[54]

In the 1975 Browns Ferry incident, when both units had emergency core cooling system failures as a result of a fire which burned out of control for hours, officials who should have been preparing to evacuate the area were not told of the danger. The civil defense coordinator heard about the fire two days after it occurred; the local sheriff said he was "asked to keep quiet about the incident to avoid any panic". Fifteen minutes elapsed between the start of the fire and when the alarm was called in; there was confusion as to whom to call.[55] In the partial core melt in the Detroit fast breeder reactor in 1966, much the same happened. No public alert was given, not because there was no danger, but instead because Detroit Edison feared 'mass panic'.[56]

Likewise in this country's worst nuclear accident, which occurred in 1979 at the Three Mile Island Generating Station in Pennsylvania, authorities were not notified of the hazardous situation until hours after it had occurred; when officials were told, they delayed giving orders for evacuation because they feared mass panic and diminishing of the image of the nuclear industry. So much confusion and cover-up were revealed in the Pennsylvania incident that a special investigative team had to be appointed by the President to assess what had happened and how the NRC had responded to the crisis. The team reported that the crisis was generated because the utility involved, the nuclear industry in general, and the Nuclear Regulatory Commission (NRC) all had minimized the danger and had made "the assumption . . . that nuclear power plants produce no risks for communities surrounding the plant".[57] "Plant officials 'forgot' to mention" one radioactive "release of unknown magnitude", lied to government officials about the seriousness of the situation, and both they and the government (NRC) officials "isolated" everyone else from accurate radiological information.[58] Because of what the commission called a "communication problem", the government and the industry "left local communities unprepared to deal with such accidents".[59] Thus state and local officials expressed "a lack of confidence in the emergency preparedness capabilities".[60] As TMI illustrated, like the effectiveness of the emergency core cooling system, the effectiveness of emergency response plants is contingent upon human intelligence, honesty, and good will. Human error can be the Achilles' heel in emergency activities.

In the light of all these occurrences and the methodological assumptions undergirding the thesis of 90% evacuation, there are strong reasons to doubt (A_4). Moreover even if the reasoning underlying the assumption of 90% evacuation were valid, there remain cogent grounds for believing that persons might not be warned of the need to evacuate. Hence it is not surprising that the American Physical Society criticized thesis (A_4). But if (A_4) is doubtful, then (A) is also questionable, and the probability of a nuclear catastrophe may not be low. But if this is the case, then there is reason to question the liability limits set by the Price-Anderson Act.

2.3. ETHICAL PROBLEMS IN THE PRICE-ANDERSON ACT

Besides the logical and methodological problems inherent in the reasoning used to justify the Price-Anderson Act, there are a number of ethical difficulties with the legislation. These are crucial to the debate over atomic energy, because both proponents and opponents of atomic power admit that nuclear fission is a very dangerous technology. The real difference between them focuses not only on (1) *how much* of a danger it poses, but also on (2) *how acceptable* this danger is when the risks are weighed against the benefits. Even if the preceding discussion about (1) revealed only sound methodological presuppositions supporting the liability limit, it would still be necessary to examine (2), the acceptability of assuming an admittedly-low risk. This examination is necessary, not only because, in the past, humankind has had to bear the consequences of 'impossible' accidents (such as the sinking of the Titanic, the falling of the Tacoma Narrows Bridge, the colliding of two planes over the Grand Canyon, and the bomber running into the Empire State Building), or because statistically an accident is just as likely to happen 'sooner' as 'later'. Rather, it is necessary to examine the acceptability of the risk, because the failure to do so will only increase the probability that society will not see nuclear power as an ethical issue. If the greatest sin is to be conscious of none, then the greatest error of 'atomic energy morality' is to believe there is none, in other words, to assume that nuclear technology raises only technological but not ethical questions.

2.3.1. *The Assumption That Nuclear Power Is Only a Technological Issue*

To some extent the formulation and governmental application of the Price-Anderson Act is built exactly on this presupposition, viz., that questions related to nuclear power are technological and not ethical ones. There is evidence that (A) and (A₃) have been taken, by government regulators of the nuclear industry, as sufficient justification for the Price-Anderson Act, and hence that they view the law as requiring only technological substantiation. So far as I know, no arguments have been made, either in the AEC Staff Study of the Price-Anderson Act, or in the Rasmussen Report (WASH-1400), to show that the Price-Anderson Act is *ethically* justifiable. Rather the authors repeatedly used theses (A) and (A₃) to 'justify' the liability limit,[61] even though (A) and (A₃) are purely technological or scientific statements.

When citizens brought lawsuits and complained that the risk of nuclear power was unfair, that thesis (A₃) was doubtful because the emergency core cooling system had not been tested, and consequently that thesis (A), regarding the impossibility of catastrophic damages, was questionable, they raised an ethical question. This was, of course, whether there ought to be a limitation on liability. The response of the courts has been to 'defer' to the 'expertise' of the Nuclear Regulatory Commission to assess the probability of an accident in which liability would exceed the Price-Anderson limit.[62]

With only one exception, the courts have viewed the liability limitation as a technical, rather than as an ethical, issue. In 1977, the US District Court in Charlotte, North Carolina, declared the Price-Anderson Act unconstitutional on the grounds that it would allow "the destruction of the property or the lives of those affected by nuclear catastrophe without reasonable certainty that the victims would be reasonably compensated". Although this court ruled that the act is in violation of the equal protection and due process provisions of the Fifth Amendment, because it places the costs for employing nuclear power "on an arbitrarily chosen segment of society, those injured by nuclear catastrophe", the ruling was reversed on June 26, 1978 by the US Supreme Court. The highest court ruled that, "when appraised in light of the extremely remote possibility of an accident in which liability would exceed the statutory limit . . . , a $ 560 million ceiling is within permissible limits and not violative of due process".[63]

This response of the courts has been predicated on the assumption that the safety issue and hence the liability issue are merely technological, and not ethical, problems. Otherwise there would have been no reason to defer to the NRC. The NRC, however, is no more competent than the individual citizen to decide if a risk is worth it. The technical question is what the risk is, not whether it should be taken, and the latter issue is the one raised by citizens in numerous 'moratorium' lawsuits based on the limits of the Price-Anderson Act. Moreover, deference to the agency which is the major proponent of both nuclear power and the Price-Anderson Act virtually denies the public a hearing on ethical issues they have raised.

Generic hearings provide another example of how government regulators view presuppositions about safety, used to justify the Price-Anderson Act, as purely technical issues. Generic issues are policy questions which arise in every licensing case, but which are not completely covered by regulation. Since such questions are not clearly decidable, the Atomic Energy Commission, and later the Nuclear Regulatory Commission, decided to remove them from individual licensing cases and treat them instead at a single, national, generic hearing, e.g., on the effectiveness of the emergency core cooling system. In this way, explains the NRC, the government need not treat "wide-ranging discussion of the nation's energy policy".[64] Instead it can "eliminate where possible from the public debate over nuclear energy extraneous arguments which cloud and make meaningful dialogue impossible".[65] Questions of safety, and whether a risk ought to be taken, are now disallowed at individual hearings regarding licensing a power plant. For all practical purposes, the ethical issues have been removed from grass-roots-level discussions, and deferred to a national forum where only technical experts are allowed to speak. Thus the value question of whether a certain group of people ought to take the chance, that the emergency core cooling system will prevent nuclear damages from exceeding the Price-Anderson limit, is thereby changed into a technical question, viz., whether the emergency core cooling system will work. This means that 'the naturalistic fallacy' has been committed.

The naturalistic fallacy consists of the attempt to deduce ethical conclusions from purely empirical premises or to define ethical characteristics in nonethical terms.[66] It is built on the false assumption that "ethics is an empirical or positive science".[67] Of course it is not, since no amount of solely factual or scientific data about *what is the case* is alone sufficient to determine ethically *what ought to be the case*. In making this assumption, those who commit the naturalistic fallacy fall into a number of errors, the most common of which is the attempt to replace ethics with one of the natural sciences.[68]

In the case of requiring generic hearings, public policy regarding nuclear power embodies the naturalistic fallacy on two counts. First, when questions relevant to ethics are disallowed at the local level, and said to be 'generic', this means that only purely technical questions will be treated at the individual plant licensing hearing which determines whether a specific group of people will have a nuclear reactor or not. The second commission of the fallacy occurs when, at the national generic hearings on a particular safety topic, only technical experts are allowed to take part.

As has been pointed out, citizens are able to participate only in the technical aspects of the hearings.[69] Hence theses such as (A_3), that the emergency core cooling system is reliable, and other purely technical propositions, are taken as sufficient grounds for asserting (A) and for accepting the Price-Anderson Act. Since "there is no forum for the public to express its views on the policy questions involved",[70] public policy or ethical questions are simply ignored.

The naturalistic fallacy is evident not only in the types of evidence acceptable in a court or hearing relevant to the liability limit, but also in the evidence cited in government studies. For example, the authors of the only allegedly complete government study on nuclear power (WASH-1400, the Rasmussen Report) explain that a nuclear accident would cause 5000 deaths, in addition to cancers and genetic damage. Next, instead of arguing why this risk is moral, they write: "however such injuries would be insignificant compared to the eight million injuries caused annually by other accidents . . . the small increases in these diseases would not be detected".[71] They also argue that accidents such as fires, air crashes, etc., are more probable than nuclear disasters.[72] Likewise, in a recent government publication, the Energy Research and Development Agency argued for the thesis that nuclear plants were safe (and therefore that the insurance limitation was reasonable) by claiming: "predicted nuclear accident risks are small compared to other possible causes of fatal injuries".[73] The purpose of their arguments has been to show why extensive damages from nuclear accidents are unlikely, and hence why the Price-Anderson Act is an acceptable law. Showing that damages resulting from taking a risk are low, undetectable, or less probable than those accruing from other risks, however, does not provide reasons why the risk is moral or ought to be taken. As a number of contemporary philosophers have pointed out, one cannot deduce a normative or ethical statement from those which are solely descriptive or scientific, without committing the naturalistic fallacy.[74] As Albert Einstein put it: "Scientific statements of facts and relations, indeed, cannot produce ethical directives".[75] (For further information on the naturalistic fallacy and its commission in assessments of nuclear technology, see Chapter Six.)

2.3.2. *Utilitarian Distributions of Nuclear Costs and Benefits*

If it were possible to produce ethical directives solely on the basis of scientific facts, then it would be moral to do whatever it was scientifically possible to do. However this might violate considerations of equity. For example, if something (generation of electricity by means of nuclear energy) *benefits* a given set of people, why is it equitable for only a subset of that group to bear the *costs* (inadequate compensation for damages caused by a nuclear accident) of that activity? Considerations of equity demand that the rights of the minority cannot be ignored in order to benefit the majority. If they are forgotten, then the very concept of right becomes meaningless. What is operative instead is a concept of utility.[76]

While it is realistic to recognize that costs and benefits of technology can never be divided completely equitably, ethics demands that this be done as far as possible. Hence to deny, for reasons of financial expediency, the rights of a minority to collect damages owed them, is inequitable, unjust, and arbitrary. Such a liability limit is arbitrary because it discriminates against people unfortunate enough to be living in the environs of a nuclear plant. The Fifth and Fourteenth Amendments, on the other hand, apply to all persons in the United States, and do not allow discrimination on the basis of geographical considerations. They forbid depriving "any person of life, liberty, or property, without due process of law".

What the Price-Anderson Act allows, by virtue of the limit on
liability, is the situation where the public can be damaged, yet
where the public cannot recover damages. For this reason it is
contrary to equal justice, due process, and equal protection under
the law.[77]

The limitation on liability of the Price-Anderson Act violates
equal justice in ways other than its neglect of the rights of those
living near a nuclear reactor. Because the act is governed by a 20-
year statute of limitations, and since radiation-induced cancers
often take thirty years before symptoms are apparent,[78] the act
also discriminates against injured parties whose symptoms do not
appear as early as those of others. This means that a disproportionate share of the risk of Price-Anderson is, again, borne by
victims.

To the preceding arguments regarding the utilitarian emphasis
of the Price-Anderson Act's limit on liability, many government
regulators have responded that the limitation does not preclude
the possibility that, were there a disaster, Congress would extend
funds to the victims in excess of the limit specified.[79] The ethical
problem with this response is, of course, that it does not guarantee
rights to property, including rights to collect damages. If the
distinction between 'utility' and 'right' means anything, it is not
simply that harm will *in fact* be compensated, but that harm will
in principle be compensated. Moreover, the pragmatic difficulty
with this response is that, unless it is required by law or obtained
through the courts (which Price-Anderson excludes), compensation
for damages is almost certain not to attain 100%.

2.3.3. *The Threshold Clause and the Argument from Ignorance*

The provisions of the Price-Anderson Act threaten authentic
morality, however, for reasons other than the fact that its liability
limit exchanges a system of due process, property rights, and
equal rights for a system of utility. The act also allows industry

representatives to use the argument from ignorance whenever nuclear accident victims sue for damages. Let us see why this is the case.

The Price-Anderson Act provides that damages are collectible *only when* the Nuclear Regulatory Commission certifies that an "extraordinary nuclear occurrence" has taken place. Since there is no such thing as a radiation threshold below which there is no deleterious biological effect, and therefore no safe dose of radiation,[80] use of such a 'threshold rule' virtually necessitates that some cancer victims of a nuclear accident will go uncompensated.[81] What might occur is precisely what has happened in the past. When the partial core melt of the Fermi fast breeder reactor caused a major release of radiation in 1966 in Detroit, the public was simply told that the heightened radiation levels presented no health threat.[82] In 1961, when the explosion and partial core melt of the Idaho Falls reactor killed three persons, the radiation alarm sounded and signalled releases of radiation into the atmosphere outside the building,[83] but government officials said that these releases were not immediately harmful to the public. Likewise, in the 1957 Windscale (England) nuclear accident, a burst uranium fuel element caused a uranium fire and a massive release of radiation into the atmosphere. As a consequence, radioactive iodine deposited on pastures and foliage posed a threat to infants and children in the area. Radiation levels were so high that cows' milk from an area covering 200 square miles was confiscated. Fallout from the accident reached London, which was 300 miles away. Yet workers at the reactor and people from the surrounding area were told that there was "no danger to the public". Moreover government officials lied when they claimed that the harmful radiation had been carried out to sea.[84] Likewise when the 1975 nuclear accident occurred at Brown's Ferry, Alabama, the public was assured that there had been no release of harmful levels of radiation. The difficulty with this claim, however, is that the radiation monitors in the stacks were not working.[85]

Much the same situation occurred in the 1979 Three-Mile-Island (TMI) accident. Although federal officials have repeatedly assured the public that "certainly not more than one or two persons could die as a result of that accident",[86] the US government has little basis for that claim, because it knows neither the amounts of all radioactive releases from TMI nor the exposure levels to the public. As one scientist claimed at Congressional hearings recently: there were

some rather large releases of radioactive materials [at TMI], but unfortunately, there were so few detectors out in the field that it was very difficult to tell what happened to those radioactive materials, whether they did impact on people, or whether they simply blew off and became sufficiently diluted . . . there is one dosimeter between the plant and Lancaster [Pennsylvania], and it has been that way approximately since March 28 [1979].[87]

Despite the fact that higher levels of radiation releases were documented in all these nuclear accidents, except the one at Brown's Ferry, the federal government continued to maintain, at least before the Three-Mile-Island accident, that there have been "no nuclear accidents to date".[88] Government regulators make this surprising remark because the context of their discussion is the safety of *commercial* nuclear reactors. Since most reactors have been military, experimental, or government-subsidized, they are not classed purely as 'commercial', and accidents in them are not called "accidents of commercial reactors". Government (Nuclear Regulatory Commission) officials were reluctant to call the TMI incident an "accident", because of "the damage this could do the [nuclear] industry's image".[89] Hence if future accidents may be defined as 'nonaccidents', and if abnormally high releases of radiation may be called 'safe', then there is no assurance that the NRC will accurately define an 'extraordinary nuclear occurrence'. And, of course, without this definition no

damages may be collected under Price-Anderson provisions. Therefore there is reason to believe that, in the case of many accidents,
the threshold provision of the act would function as a utilitarian
principle according to which some victims of nuclear injuries might
not be compensated. Moreover because the act requires payment
to be given only "upon proof of the causal relationship between
the occurrence and the injury . . . and upon proof of damage",[90]
it is likely that some injured persons will be unable to prove this
and hence likely that they will remain uncompensated.

Although proof of injury is not difficult for the more serious
and immediate types of radiation deaths and injuries, it is almost
impossible in cases of cancers and genetic effects whose symptoms
are manifested only years later. How could one *prove*, for example,
that a cancer (appearing 15 years after a person's exposure to
radiation released in a nuclear accident) was truly caused by the
accident?

The difficulty of proving that radiation has caused a particular
cancer is illustrated very well by recent attempts of US veterans
to prove that their cancers were caused by their participation in
atomic bomb testing in the 1950's. On August 31, 1957, for
example, US troops were marched "to within 300 yards of ground
zero" soon after the Smoky atomic blast. Although the US Department of Energy has found records of 900 over-exposed individuals
involved in tests dating from 1951–1962, benefits have been
granted only to ten men of the many "who claimed their disabili-
ties were the result of radiation exposure received during the
US nuclear weapons tests".[91] Because of the problems associated
with all levels of radiation and the possible violations of rights
involved, the US House Commerce Subcommittee on Health and
Environment held three days of hearings in January, 1978, to hear
testimony by servicemen.[92] According to Orville Kelly, USN
Commander of Eniwetok Atoll during weapons tests in the 50's,
between 200 and 400 thousand soldiers were exposed to high levels

of radiation. Since radiation was not completely monitored, he says, the government claims it has records of only 900 overexposures. Kelly, a cancer victim and founder of 'Make Today Count', has been denied benefits because the Veterans Administration says he cannot prove his cancer was caused by radiation received during the weapons tests. Kelly was within 5 miles of ground zero during 23 nuclear tests and was exposed to fallout. In the last seven years of pleading for benefits for his wife and four children, he has exhausted all legal channels of appeals.[93]

The case of servicemen exposed to fallout in the fifties poses many problems similar to those faced by citizens attempting to collect damages under the Price-Anderson Act. If 25% of the population will contract cancer, apart from nuclear reactor hazards, how would one distinguish these incidences from the malignancies resulting from a power plant accident? The ethical dilemma here is not just that some people who deserve to be compensated will not be, because of the 'proof' and 'threshold' clauses of the Act, but also that, *in principle*, there is no way to prove whether or not radiation from a particular plant has caused a cancer. Hence retaining these clauses forces government and the courts to employ an *argumentum ad ignorantiam*, an argument from ignorance. They will have to conclude that an accident did not cause a particular cancer whenever the victim is unable to prove that it did. This means that employment of the Price-Anderson Act puts a double burden on citizens. It has not only placed utility interests above those of consumers in limiting due process and equal protection under the law, but also (because of the very nature of the radiation hazard) made it impossible, *in principle*, to adjudicate radiation-injury claims involving equal justice under law.

In the event that no nuclear accident occurs, of course, the Price-Anderson Act provides important benefits. It lessens utility costs for nuclear generation of electricity and therefore ultimately enables customers to pay lower prices for the energy they use. Moreover it can be argued that both these assests contribute to the economic and industrial health of the country and therefore to the well-being of many US citizens. The price paid for these benefits, however, is (1) the acceptance of a utilitarian calculus for distributing costs and benefits of energy generation and consumption; (2) the assumption that a nuclear catastrophe will not occur; (3) the possible violation of rights to equal justice and due process; and (4) the placement of financial responsibility for a possible reactor accident upon the shoulders of its victims rather than its perpetrators.

3. Conclusion

Admittedly the problems with the Price-Anderson Act do not allow one to conclude that this country ought not to use nuclear fission to generate electricity. At best, this analysis indicates only that current public policy regarding nuclear liability assigns a higher priority to providing inexpensive electricity and protecting utility interests than to preserving Constitutional guarantees of due process and equal protection. To the extent that protection of basic rights is a necessary condition for ethical use of any technology, then to that degree, the Price-Anderson Act ought to be assessed very carefully. If one opposes repeal of this legislation, let it not be for the same reason that New England residents, whose views were described in the diary of Ralph Waldo Emerson, opposed construction of lighthouses on Cape Cod. Although the lights were badly needed as a protection for mariners, the people made their living in the shipwreck business. They found it profitable not to provide such protection.

Notes

[1] Decimus Junius Juvenal, *Satires* **VI** (1), 292. Interestingly, however, Plato thought it was absurd to think that a guardian should himself need a guardian. See *The Republic* **III**, 403E.

[2] Other writers have made a similar point. See Garrett Hardin, 'The Tragedy

of the Commons', in *Technology and Society* (ed. by N. de Nevers), Addison-Wesley, Reading, 1972, p. 164; and D.J. Rose, 'New Laboratories for Old', in *Science and Its Public: The Changing Relationship* (ed. by G. Holton and W. Blanpied), D. Reidel Publ. Co., Dordrecht, 1976, pp. 151–52.

[3] Alasdair MacIntyre, 'Utilitarianism and Cost-Benefit Analysis: An Essay on the Relevance of Moral Philosophy to Bureaucratic Theory', in *Values in the Electric Power Industry* (ed. by K.M. Sayre), University Press, Notre Dame, 1977, pp. 217–18; hereafter cited as: *Values.*

[4] Schweitzer, *The Philosophy of Civilization*, cited by J.L. Huffman, 'Individual Liberty and Environmental Regulation: Can We Protect People While Preserving the Environment?', *Environmental Law* VII (3), (Spring 1977), 447.

[5] See Chapter One, Section 1, 'The History of Nuclear Energy'.

[6] Industry's refusal to invest in nuclear technology is documented in Chapter One, Section 2, 'Government Regulation of Atomic Power', and in Section 1.

[7] J.R. Brydon, 'Slaying the Nuclear Giants', *Pacific Law Journal* VIII (2), (July 1977), 781.

[8] 'AEC Staff Study of the Price-Anderson Act, Part I', *Atomic Energy Law Journal* XVI (3), (Fall 1974), 220.

[9] See Chapter One, Sections 1 and 2. They document industry's insistence that a government-guaranteed liability limit is a necessary condition for their employment of nuclear technology.

[10] For discussion of the temporary nature of the act, see Chapter One, Section 2.

[11] J. Marrone, 'The Price-Anderson Act: The Insurance Industry's View', *Forum* XII (2), (Winter 1977), 608–609; hereafter cited as: 'Price-Anderson'.

[12] For discussion of this point, see Chapter One, Section 2, especially documentation of the reasons behind the Energy Reorganization Act of 1975.

[13] For bibliographical material regarding this amendment, see note 14.

[14] Explanations of these two amendments to the Price-Anderson Act can be found in R. Lowenstein, 'The Price-Anderson Act', *Forum* XII (2), (Winter 1977), 599–603; hereafter cited as: 'P-AA'. See also Marrone, 'Price-Anderson', pp. 608–609; M. McCormack, 'U.S. Congressional Attitudes and Policies Affecting Nuclear Power Development in the World', *Atomic Energy Law Journal* XVII (4), (Winter 1974), 307–309; R. Wilson, 'Nuclear Liability and the Price-Anderson Act', *Forum* XII (2), (Winter 1977), 613–14; F. Schmidt and D. Bodansky, *The Energy Controversy: The Fight over Nuclear Power*, Albion, San Francisco, 1976, 144; hereafter cited as: *Energy Controversy.*

[15] W.S. Caldwell *et al.*, 'The "Extraordinary Nuclear Occurrence" Threshold and Uncompensated Injury under the Price-Anderson Act', *Rutgers-Camden Law Journal* VI (2), (Fall 1974), 360; hereafter cited as: 'Uncompensated Injury'.

[16] 'AEC Staff Study', p. 231.

[17] Failure of the technological controls in a reactor would result in disastrous consequences. "A million-kilowatt nuclear power plant, after a year of operation, contains ten billion curies of radioactive material, enough (if properly distributed) to kill every one in the U.S." (Sheldon Novick, *The Electric War*, Sierra, San Francisco, 1976, pp. 152–53.) Moreover much of this radiation remains lethal (capable of ionization causing death, cancer and genetic damage) for hundreds of thousands of years. Hence any extensive contamination with plutonium 239 is considered by scientists to be 'permanent'. In WASH-740, the Brookhaven Report, the (pro-nuclear) government arrived at conservative estimates for effects of a core melt, and resultant releases of radioactivity. The report concluded that such an accident could lead to 45,000 immediate deaths, 100,000 injuries, innumerable cancers, contamination of an area the size of Pennsylvania and up to $17 billion in property damage. (See 'Theoretical Possibilities and Consequences of Major Accidents in Large Nuclear Power Plants', USAEC Report WASH-740, Government Printing Office, Washington, D.C., 1957, and its update, i.e., R.J. Mulvihill, D.R. Arnold, C.E. Bloomquist, and B. Epstein, 'Analysis of United States Power Reactor Accident Probability', PRC R-695, Planning Research Corporation, Los Angeles, 1965. See also J. Elder, 'Nuclear Torts: The Price-Anderson Act and the Potential for Uncompensated Injury', *New England New Review* XI (1), (Fall 1975), 127; T. Black, 'Population Criteria of Oregon and the U.S. in the Siting of Nuclear Power Plants', *Environmental Law* VI (3), (Spring 1976), 897; J.J. Berger, *Nuclear Power*, Dell, New York, 1977, p. 45.)

[18] J. G. Fuller, *We Almost Lost Detroit*, Ballantine, New York, 1975, p. 60; hereafter cited as: *Detroit*.

[19] Richard Wilson, 'Nuclear Liability and the Price-Anderson Act', *Forum* XII (2), (Winter 1977), 617, who is a proponent of nuclear power and of the Price-Anderson Act, provides this statistic about airline coverage.

[20] The reason given by the Atomic Energy Commission for the necessity of government-subsidized insurance for the nuclear industry is that private insurance would not take the risk. See 'AEC Staff Study of the Price-Anderson Act, Part I', pp. 219–20.

[21] This is a per-year, per-reactor probability, computed by Senator Mike Gravel (Alaska), and quoted in J.J. Berger, *Nuclear Power*, Dell, New York, 1977, pp. 145–46.

[22] U.S. Nuclear Regulatory Commission, *Reactor Safety Study – An Assessment of Accident Risks in U.S. Commercial Nuclear Power Plants*, Report No. (NUREG-75/014) WASH-1400, Goverment Printing Office, Washington, 1975, pp. 157 ff.; hereafter cited as: 'WASH-1400'. This figure is also a per-year, per-reactor probability.

[23] Former Insurance Commissioner of the State of Pennsylvania, H.S. Denenberg, calculated this annual premium; he is quoted in M.C. Olson,

Unacceptable Risk, Bantam, New York, 1976, p. 56. See also H.S. Denenberg, *Citizen's Bill of Rights and Consumer's Guide to Nuclear Power*, Pennsylvania Insurance Department, Reading, 1973, pp. 1–5. For information on the $ 5 million maximum deferred premium established by the Price-Anderson Act, see Lowenstein, 'P-AA', pp. 599–603; Marrone, 'Price-Anderson', pp. 608– 609; Wilson, 'Nuclear Liability', pp. 613–14.

[24] See Chapter One, Sections 1 and 2.

[25] Several authors have made points related to this argument of mine; see J.W. Gofman and A. Tamplin, *Poisoned Power*, Rodale, Emmaus, Pennsylvania, 1971, p. 357; and R. Nader, 'Nuclear Power: More than a Technological Issue', *Mechanical Engineering* **XCVIII** (2), (February 1976), 33.

[26] See Wilson, 'Nuclear Liability', pp. 617–18; and Schmidt and Bodansky, *Energy Controversy*, p. 144.

[27] M. McCormack, 'U.S. Congressional Attitudes and Policies Affecting Nuclear Power Development in the World', *Atomic Energy Law Journal* **XVII** (4), (Winter 1974), 311.

[28] R. Drinan, 'Nuclear Power and the Role of Congress', *Environmental Affairs* **IV** (4), (Fall 1975), 607. See also Gravel, cited by Gofman and Tamplin, *Poisoned Power*, p. 364.

[29] Berger, *Nuclear Power*, pp. 112, 144–47.

[30] Gofman and Tamplin, *Poisoned Power*, p. 362.

[31] NRC, 'WASH-1400', pp. 157, 195, 223, 226.

[32] See 'AEC Staff Study of the Price-Anderson Act, Part II', p. 297, where the Atomic Energy Commission makes the argument that the likelihood of a serious accident is remote.

[33] See NRC, 'WASH-1400'. See also US AEC, 'Reactor Safety Study', *Atomic Energy Law Journal* **XVI** (3), (Fall 1974), 177–204; hereafter cited as: 'Reactor Safety'.

[34] Report of the APS cited in R. Doctor *et al.*, 'The California Nuclear Safeguards Initiative', *Sierra Club Bulletin* **LXI** (5), (May 1976), 45; hereafter cited as: 'Safeguards'. The APS critique may be found in H.W. Lewis *et al.*, 'Report to the American Physical Society by the Study Group on Light-Water Reactor Safety', *Reviews of Modern Physics* **XLVII** (1), (Summer 1975), SI–S124; and Study Group on Light Water Reactor Safety, 'Nuclear Reactor Safety – the APS Submits Its Report', *Physics Today* **XXVIII** (7), (July 1975), 38–43. See also Schmidt and Bodansky, *Energy Controversy*, pp. 86–87. The APS critique is given in NRC, 'WASH-1400', Appendix XI, pp. 2–8ff.

[35] Cited in R. Augustine, 'NRC Finds Safety in Numbers', in *Countdown to a Nuclear Moratorium* (ed. by R. Munson), Environmental Action Foundation, Washington, D.C., 1976, p. 31. See also Berger, *Nuclear Power*, pp. 54–63. The Union of Concerned Scientists makes these same criticisms in NRC, 'WASH-1400', Appendix XI, p. 2-12; see also pp. 94–103 and Appendix I.

[36] Nader, 'Nuclear Power', pp. 32–33, and Novick, *Electric War*, pp. 156–58, 315–16.

[37] C. Hohenemser, R. Kasperson, and R. Kates, 'The Distrust of Nuclear Power', *Science* CXCVI (4285), (April 1977), 25–34.

[38] Norman Rasmussen, 'Methods of Hazard Analysis', in *The Three Mile Island Nuclear Accident*, The New York Academy of Sciences, New York, 1981, p. 29.

[39] P (an annual TMI-type accident in at least 1 of 151 reactors now operating or under construction in the US) = 1 − P (no annual TMI-type accident in the US) = $1 - (1 - (1/250))^{151} = 1 - (0.996)^{151} = 1 - 0.545 = 0.455$ or 46%.

[40] Nuclear proponents such as Schmidt and Bodansky, *Energy Controversy*, p. 143, make it clear that the Rasmussen Report has been and ought to be used as a justification for the Price-Anderson Act.

[41] See Black, 'Population Criteria', p. 897; Elder, 'Nuclear Torts', p. 127; Berger, 'Nuclear Power', pp. 44–46; Gofman and Tamplin, *Poisoned Power*, pp. 174–79, 354–55; and G. C. Coggins, 'The Environmentalist's View of AEC's "Judicial" Function', *Atomic Energy Law Journal* XV (3), (Fall 1973), 189. Elder, 'Nuclear Torts', p. 131; and J.G. Palfrey, 'Energy and the Environment', *Columbia Law Review* 74 (8), (December 1974), 1398–1400, also reveal the 'cover-up' of the Brookhaven Report and of problems with the emergency core cooling system.

[42] See Barry Commoner, *The Closing Circle*, Bantam, New York, 1974, p. 197.

[43] W. O. Doub, 'Meeting the Challenge to Nuclear Energy Head-On', *Atomic Energy Law Journal* XV (4), (Winter 1974), 258–59; hereafter cited as: 'Challenge'.

[44] Olson, *Unacceptable Risk*, pp. 72, 86, 82. See also C. L. McGuire, 'Emerging State Programs To Protect the Environment: Little NEPA's and Beyond', *Environmental Affairs* V (3), (Summer 1976), 572–75; Palfrey, 'Energy', p. 1378; Coggins, 'Environmentalist's View', pp. 182–84; and J. Bieber, 'Calvert Cliffs Coordinating Committee v. AEC: The AEC Learns the True Meaning of the National Environmental Policy Act of 1969', *Environmental Law* III (2), (Summer 1973), 316–33.

[45] Palfrey, 'Energy', p. 1392.

[46] Gofman and Tamplin, *Poisoned Power*, pp. 225–26.

[47] See Coggins, 'Environmentalist's View', pp. 184, 187; and Novick, *Electric War*, pp. 109, 256 for details of harrassment and firing of government scientists who refused to allow pro-industry censorship. In Gofman and Tamplin, *Poisoned Power*, esp. pp. 24–27, 135, 255–72, the authors (former AEC scientists) tell how they were commissioned to calculate radiation effects, but when they arrived at conclusions contrary to the thesis that currently allowed radiation levels specified by the AEC (NRC) were safe, they were fired and the results of their studies were suppressed.

[48]   V. McKim, 'Social and Environmental Values in Power Plant Licensing',
in Sayre, *Values*, pp. 46–54, tells of interviewing Argonne staff members who
said that top-level Nuclear Regulatory Commissioners had tampered with
environmental impact statements.

[49]   NRC, 'WASH-1400', p. 164; and Novick, *Electric War*, p. 155. See also
J. Primack and F. Von Hippel, 'Nuclear Reactor Safety', *The Bulletin of the
Atomic Scientists* **XXX** (8), (October 1974), 7–9; and M. Bauser, 'United
States Nuclear Export Policy: Developing the Peaceful Atom as a Commodity
in International Trade', *Harvard International Law Journal* **XVIII** (2), (Spring
1977), 51.

[50]   Primack and Von Hippel, 'Nuclear Reactor Safety', pp. 7 and 9; Berger,
Nuclear Power, pp. 46–52.

[51]   Both proponents and opponents of nuclear power admit that this reason
for failing to make the scale-model tests is the correct one. See Schmidt and
Bodansky, *Energy Controversy*, pp. 139–142; and Novick, *Electric War*,
p. 192.

[52]   D. M. Rubin, *et al.*, *Report of the Public's Right to Information Task
Force*, The President's Commission on the Accident at Three Mile Island,
US Government Printing Office, Washington, DC, 1979, p. 7; hereafter cited
as: Commission, *Information*. For additional information on the TMI accident,
see notes 38, 39, 57–60, and 86–89 in this chapter, as well as: Comptroller
General of the US, *Three Mile Island: The Most Studied Nuclear Accident
in History*, US Government Printing Office, Washington, DC, 1980; Subcommittee on Energy and the Environment of the Committee on the Interior
and Insular Affairs, *Accident at the Three Mile Island Nuclear Powerplant*,
Oversight Hearings, US House of Representatives, 96 Congress, US Government Printing Office, Washington, DC, 1979; and Subcommittee on Energy,
Nuclear Proliferation and Federal Services of the Committee on Governmental Affairs of the United States Senate, *Impact Abroad of the Accident
at the Three Mile Island Nuclear Power Plant: March–September 1979*, US
Government Printing Office, Washington, DC, 1980 and Hearing Before the
Subcommittee on Energy Research and Production of the Committee on
Science and Technology, US House of Representatives, *Kemeny Commission
Findings*, US Government Printing Office, Washington, DC, 1979, esp.
statement by J. G. Kemeny, pp. 5–10.

[53]   The 90% assumption is explicitly made in NRC, 'WASH-1400', p. 111,
despite the fact that evacuations would need to cover approximately 400
square miles ('WASH-1400', p. 234) and that the maximum warning time
prior to a disaster would be 2½ hours ('WASH-1400', p. 118). See also
Doctor *et al.*, 'Safeguards', p. 45.

[54]   R. Polluck, 'Nuclear Emergency Plans Fail in Most States', *Critical Mass
Journal* **III** (1), (February 1978), 1, 8.

[55] D. D. Comey, 'The Brown's Ferry Incident', in Munson, *Countdown*, pp. 5–7. See also Berger, *Nuclear Power*, pp. 40–43.

[56] Fuller, *Detroit*, p. 2.

[57] R. R. Dynes, *et al.*, *Report of the Emergency Preparedness and Response Task Force*, The President's Commission on the Accident at Three Mile Island, US Government Printing Office, Washington, DC, 1979, p. 96. For analyses of the TMI accident, see note 52 above and Report of the President's Commission, *The Accident at Three Mile Island*, US Government Printing Office, Washington, DC, as well as Committee on Interior and Insular Affairs, *Financial Implications of the Accident at Three Mile Island*, Oversight Hearings, US House of Representatives, US Government Printing Office, Washington, DC, 1981.

[58] Dynes, *Report* (note 57), pp. 91, 96. See also Committee on Interior and Insular Affairs, *Reporting of Information Concerning the Accident at Three Mile Island*, US Government Printing Office, Washington, DC, 1981, p. 93.

[59] Dynes, *Report* (note 57), p. 96.

[60] Military Installations and Facilities Subcommittee of the Committee on Armed Services, US House of Representatives, *Civil Defense and the Three Mile Island Nuclear Accident*, US Government Printing Office, Washington, DC, p. 3.

[61] See notes 31–35.

[62] M. S. Young, 'A Survey of the Governmental Regulation of Nuclear Power Generation', *Marquette Law Review* LIX (4), (1976), 843–44.

[63] Duke Power Company v. Carolina Environmental Study Group, Inc., *et al.* and U.S. Nuclear Regulatory Commission *et al.*, v. Carolina Environmental Study Group, Nos. 77-262 and 77-375, *The United States Law Week* XLVI (50), (June 1978), 4845; and Berger, *Nuclear Power*, pp. 145–46.

[64] E. D. Muchnicki, 'The Proper Role of the Public in Nuclear Power Plant Licensing Decisions', *Atomic Energy Law Journal* 15 (1), (Spring 1973), 42; hereafter cited as: 'Public'.

[65] Doub, 'Challenge', p. 261.

[66] G. E. Moore, *Principia Ethica*, University Press, Cambridge, 1951, p. 39; hereafter cited as: *PE*.

[67] *PE*, p. 39.

[68] *PE*, p. 40.

[69] Muchnicki, 'Public', p. 43. See also J. Lieberman, 'Generic Hearings: Preparation for the Future', *Atomic Energy Law Journal* XVI (2), (Summer 1974), 142. Here Lieberman argues that the procedural rights of participants in hearings, the nature of the issues to be resolved, and the fairness of the proceeding are the major problems not only of nuclear power plant hearings, but also of "decision making in a technical society".

[70] Muchnicki, 'Public', p. 44.

[71] AEC, 'Reactor Safety', pp. 182–83.

[72] AEC, 'Reactor Safety', p. 179; NRC, 'WASH-1400', pp. 187–221. Hans Bethe, a government consultant and proponent of nuclear energy makes the same technological/probabilistic argument. After admitting that a nuclear accident might cause 5000 cancer deaths, he argues: "one should remember that in the U.S., there are more than 300 000 deaths every year from cancers due to other causes". (Hans Bethe, 'The Necessity of Fission Power', *Scientific American* **CCXXXIV** (1), (January 1976), 26.)

[73] *How Probable Is a Nuclear Plant Accident?*, Office of Public Affairs, Washington D.C., 1976, EDM-074R (0-76), p. 2.

[74] A. R. White, *G. E. Moore, A Critical Exposition*, Basil Blackwell, Oxford, 1958, pp. 122-7; F. Snare, 'Three Sceptical Theses in Ethics', *American Philosophical Quarterly* **XIV** (2), (April 1977), 129–30; L. Kohlberg, 'From Is to Ought: How To Commit the Naturalistic Fallacy and Get Away with It in the Study of Moral Development', in *Cognitive Development and Epistemology* (ed. by T. Michel), Academic, New York, 1971, p. 154; J.H. Olthuis, *Facts, Values and Ethics*, Van Gorcum, Assen, 1969, pp. 28–34; K.E. Goodpaster (ed.), *Perspectives on Morality: Essays by William K. Frankena*, University Press, Notre Dame, 1976, pp. 16–17, 210.

[75] A. Einstein, 'The Laws of Science and the Laws of Ethics', in *Readings in the Philosophy of Science* (ed. by H. Feigl and M. Brodbeck), Appleton-Century-Crofts, New York, 1953, p. 779.

[76] For arguments against utilitarianism and in favor of egalitarianism, see Chapter Two, Section 3.1 and 3.2. K. Sayre and K. Goodpaster, 'An Ethical Analysis of Power Company Decision-Making', in Sayre, *Values*, pp. 238–88, argue that the utility industry operates according to the concept of utility, rather than according to the concept of right. If their analysis is correct, then it would not be surprising if the Price-Anderson Act, which is essential to utility operation in the nuclear area, is also utilitarian in nature.

[77] Rep. Cole made this same point during Congressional hearings on the Price-Anderson Act. His remarks are cited by Elder, 'Nuclear Torts', p. 134. The US District Court for the Western District of North Carolina also agreed with this opinion in a 1977 ruling. The court said that the liability limits of the Price-Anderson Act violated the due process and the equal protection clauses of the Fifth Amendment. This ruling, however, was overturned by the US Supreme Court (see note 56).

[78] See Olson, *Unacceptable Risk*, p. 55; see also Elder, 'Nuclear Torts', pp. 126–132; and J.C. Bodie, 'The Irradiated Plaintiff: Tort Recovery Outside Price-Anderson', *Environmental Law* **VI** (3), (Spring 1976), 859–96.

[79] For arguments to this effect, see 'AEC Staff Study of the Price-Anderson Act, Part II', p. 297; Schmidt and Bodansky, *Energy Controversy*, p. 145; Lowenstein, 'P-AA', p. 600.

[80] Standards of the International Commission on Radiation Protection, the

basis for US regulations governing atomic power, affirm that there is no safe dose of radiation, and no threshold below which some carcinogenic or mutagenic effects do not occur. ICRP standards are given in J.P. Holdren, 'Hazards of the Nuclear Fuel Cycle', *The Bulletin of the Atomic Scientists* **XXX** (8), (October 1974), 16, and confirmed in Gofman and Tamplin, *Poisoned Power*, pp. 39, 293–307; Gofman and Tamplin, 'Nuclear Power, Technology and Environmental Law', *Environmental Law* **II** (2), (Winter 1971), 59–63; B. Commoner, *The Closing Circle*, Bantam, New York, 1974, p. 194; and O. Hansen, 'Development and Application of Radiation Protection Standards', *Idaho International Law Review* **XII** (1), (Fall 1975), 1–32.

[81]   Caldwell *et al.*, 'Uncompensated Injury', pp. 380–383, also makes this point; see Elder, 'Nuclear Torts', p. 122.

[82]   Higher radiation readings in the atmosphere outside the reactor building are documented in Fuller, *Detroit*, p. 17.

[83]   Fuller, *Detroit*, pp. 109–116.

[84]   Fuller, *Detroit*, p. 84; see pp. 75–92 for more details on the Windscale accident.

[85]   See Schmidt and Bodansky, *Energy Controversy*, p. 88; and Fuller, *Detroit*, pp. 17ff.

[86]   Subcommittee on Energy Research and Production of the Committee on Science and Technology of the US House of Representative, *Nuclear Powerplant Safety After Three Mile Island*, US Government Printing Office, Washington, DC, 1980, p. 68. For additional information on the TMI accident, see note 52 in this chapter.

[87]   Dr. Chauncey Kepford before the Subcommittee on Natural Resources and Environment of the Committee on Science and Technology of the US House of Representatives, *Three Mile Island Nuclear Plant Accident*, US Government Printing Office, Washington, DC, pp. 3–4.

[88]   'AEC, Reactor Safety', p. 178. This claim was made prior to the 1979 accident at Three Mile Island, Pennsylvania. Since major releases of radiation occurred there, but no documentation of deaths caused by this incident exists, it is not clear whether the NRC would term this an 'accident' in its technical sense of the term.

[89]   Commission, *Information* (note 52), p. 8.

[90]   Cited by Caldwell *et al.*, 'Uncompensated Injury', p. 366.

[91]   M. Korchmar, 'Radiation Hearings Uncover Dust', *Critical Mass Journal* **III** (12), (March 1978), 5; see also Berger, *Nuclear Power*, pp. 65–66, 69, 71; and R. Kraus 'Environmental Carcinogenesis: Regulation on the Frontiers of Science', *Environmental Law* **VII** (1), (Fall 1976), 83–135.

[92]   Korchmar, 'Radiation Hearings', p. 5.

[93]   Facts regarding the Orville Kelly case, one of many in which benefits were denied, came from a personal conversation with him in my home on August 17, 1978.

Chapter Five

Nuclear Economics and the Problem of Externalities

The considerations of the previous chapters have made it clear that the debate over nuclear energy provides an excellent instance of the age-old tension between the ethical principle of equity and the economic principle of expediency or efficiency. The distinquished economist, Arthur Okun, describes this classic conflict as a controversy over whether democratic political systems or market relations shall govern social processes.[1]

In this chapter I will evaluate the effects of certain market relations on public policy governing atomic power. Specifically I will assess the societal impact of several externalities of the nuclear fuel cycle.

1. The Problem of Externalities

'Externalities' are social costs. Because the social costs of producing certain goods, e.g., energy, are often not calculated, many persons wrongly assume that they do not exist. In fact, classical economic theory is based on the premise that one ought to assess all the purely *private* costs and benefits, affecting only the producers and consumers of goods, but that one can in large part ignore the costs and benefits that an enterprise involuntarily imposes on the *public*. This premise is widely accepted because "there are costs or benefits which accrue to society, but which are not included" in the original contract between parties to an exchange.[2] These costs and benefits which are not accounted for are called 'disamenities', 'diseconomies', 'spillovers', or 'externalities' since they are items external to the standard cost-benefit calculus used to justify an exchange. Nonbiodegradable pesticides, for example, are negative externalities; their health costs are borne largely by the public, but their economic benefits are received primarily by their producers and crop owners.[3]

In theory externalities may be either positive or negative, although it is easier to find examples of the latter type. Like negative externalities, positive ones are also involuntarily imposed; unlike negative externalities, positive externalities are economically beneficial to those upon whom they are placed. Such an externality is enjoyed, for example, by a homeowner who lives next to a quiet, spacious park.

1.1. THE FAILURE TO ASSESS EXTERNALITIES

Two obvious questions are raised by the presence of numerous negative externalities within the economic system. First, why have the classical laws of economics not provided a methodological framework within which to consider the social costs of producing certain goods, e.g., electrical energy? Secondly, is failure either to compensate for these externalities or to avoid them defensible on methodological or ethical grounds? Let us examine each of these queries.

There are numerous reasons why social costs, such as pollution, have not been included within most economic exchanges. First and most obviously, it has been assumed that what was the property of no one in particular was free to be used by anyone as he wished; hence privacy, quiet, clean air, and nonpolluted water were thought to be free commodities.[4] Secondly, if producers or suppliers of goods were required either to compensate society for the environmental injuries they caused or to avoid such damages, then many industries would be unable or unwilling to cover these

costs with their profits. Since a polluter who installs controls "cannot limit enjoyment of the resulting clean air to those who pay
for it", for example, he will be unable to recover the costs of his
investment and will fail voluntarily to undertake pollution controls.[5] Likewise airlines, for example, would go out of business
if they were required to provide recompense for the noise they
caused. Thirdly, even if compensation for disamenities were economically feasible, it might act as a cost barrier both to numerous
potential economic improvements and to economic growth itself.[6]
Fourthly, accounting for social or environmental damages within
an economic system of cost-benefit analyses would require solution of major epistemological and ethical difficulties. Some of
these are: knowing what the relevant losses are; determining how
they ought to be calculated, weighted, and compensated; providing for governmental regulation such that the individual liberty
of producers and consumers is not unjustifiably limited;[7] and
balancing the right to equal protection, possessed by the victims
of spillovers, with the need to operate efficiently demanded by
those responsible for such damages.[8]

In any exchange obviously it is desirable ethically that private
and social costs and benefits be equitably distributed, just as it is
desirable economically that the social rate of return exceed the
cost and therefore that the project be efficient. Since a perfect
exchange never exists, problems arise, not only regarding whether
to maximize equity or efficiency in a particular policy situation,
but also regarding which specific economic policies do in fact
maximize equity or efficiency.

In the famous debate between proponents and opponents of the
economic policy of no-growth, for example, participants disagree
not only about whether to maximize equity or efficiency, but also
about which economic policies will lead to the supremacy of one
or the other of these values. Some writers maintain that a no-
growth policy is a necessary condition for equitable distribution

of the goods of the earth,[9] while others argue that only an expanding economy can provide the jobs and the capital essential for a more equitable distribution of goods to the greatest number of people.[10] Likewise some authors claim that no-growth is a necessary condition for long-term economic health,[11] while others hold that only an expanding economy is efficient over the long-term.[12]

The second issue, raised by the presence of numerous negative externalities within the economic system, is whether failure to compensate for them or to avoid them is defensible on methodological or ethical grounds. One way of answering this question is to examine the economic policies governing specific externalities and then to analyze the epistemological and evaluational assumptions implicit in them.

There are a number of reasons why externalities in the nuclear fuel cycle present an excellent subject for such an analysis.[13] First, the economics of energy choices poses some of today's most troubling scientific, social, philosophical, and public policy questions. Secondly, decision makers of many governments have planned for nuclear energy to supply much of the world's electricity needs. US energy policy, for example, since the passage of the 1954 Atomic Energy Act, has been that nuclear power is necessary to provide "an economical and reliable basis" needed "to sustain economic growth".[14] Because of the energy crisis and the alleged economic benefits of nuclear fission, in 1973 government policymakers revealed their plan that, by the end of this century, 70% of all US electrical capacity would be nuclear.[15] Since 1973, however, this objective has come under increasing attack. Alvin Weinberg, although a respected advocate of atomic energy, has calculated that nuclear fission is not needed to sustain economic growth.[16] Similar charges, coupled with lowered capacity factors for atomic plants, a quadrupling of uranium prices, and delays followed by rising construction costs, have caused nuclear utilities to cancel over 100 plants in the last three years.[17]

Apart from whether it is economically wise or not (and that issue is still being debated), there are approximately 65 nuclear plants now operating in the US and 70 additional ones under construction. Since atomic energy is already a part of the fabric of US life, it would be desirable to examine current federal economic policies concerning the externalities of fission generation of electricity and to analyze some specific ethical assumptions implicit in them. Such an analysis will provide one way of determining, not only the current nuclear policy tradeoff between equity and efficiency, but also the general value framework presupposed in much energy decision-making. This, in turn, may throw some additional light on the economic arguments concerning the desirability of nuclear power. Not to engage in evaluation of this underlying philosophical point of view is to run a great risk, viz., that we as a society might make ourselves the victims both of bad economics and of bad philosophy. This is a real possibility, since philosophical theories often go uncriticized when they are thought to guide economic policies classed as common-sensical, necessary, or atheoretical. As a result, says Alasdair MacIntyre, such policies wrongly "enjoy an undeserved power".[18]

1.2. PARTIALLY-COMPENSATED AND UNCOMPENSATED EXTERNALITIES

In order to evaluate government policy regarding externalities of the nuclear fuel cycle, it is necessary to examine at least two distinct categories of social costs. Items in one class have, at least in part, been 'internalized' so that a government-determined price compensates for some or all of their cost to society. They will be termed 'partially-compensated externalities'. One such disamenity is the risk to the public of a catastrophic radiation accident.

Externalities in a second group, however, have not been internalized in any way; according to government policy, their price is borne by the public and is considered extraneous to the economic evaluation of nuclear energy. These spillovers will be referred to as 'uncompensated'. An example of an externality in this category is the long-term financial and medical risk caused by radioactive wastes.

2. Partially-Compensated Externalities of the Nuclear Fuel Cycle

Because one of the most important uncompensated externalities of the nuclear fuel cycle, radioactive waste, has already been discussed in Chapter Three, this chapter will address several partially-compensated externalities of the fission process. An analysis of the public policy governing this type of diseconomy reveals some interesting ethical presuppositions. It also provides several insights into whether disamenities, such as increased cancer rates, are being addressed adequately by the current operation of the economic system.

2.1. THE RISK OF CORE MELT

One of the most well-known partially-compensated externalities in the nuclear fuel cycle is the risk of a major accident releasing massive quantities of radiation. According to government calculations, if such an incident were to occur, it would be equivalent to 1000 Hiroshimas and could cause property damage alone as high as $ 17 billion, in addition to 45,000 immediate deaths, 100,000 cases of cancer, genetic damage, and other injury, and contamination of "an area the size of Pennsylvania".[19] As was pointed out in Chapter Four, however, the government has set a liability limit of $ 560 million on all nuclear accidents, so that total claims against a utility could not rise above this ceiling.[20] Since the liability limit set by the Price-Anderson Act amounts to only 3% of possible property damages, 97% of this externality could be legally uncompensated. Three reasons given by government for the liability limitation are that, without it, industry would withdraw from employment of nuclear power;[21] the probability of a catastrophic nuclear accident is low;[22] and the liability limit provides both industry and government with an economically 'affordable ceiling'.[23]

What are the ethical assumptions inherent in limiting compensation for this externality to $ 560 million? As was argued in Chapter Four, the limitation presupposes the acceptability of giving priority to utility protection and to production of inexpensive electricity, over preserving rights to due process and equal protection. Hence the very concept of 'right' is likely to become meaningless, and to be replaced by a concept of utility. As Callahan puts it, "the concept of a 'right' becomes meaningless if rights are wholly subject to tests of economic, social, or demographic utility, to be given or withheld depending upon their effectiveness in serving social goals".[24]

On the other hand, however, it is realistic to recognize that the costs and benefits of technology can never be divided completely equitably. This fact is particularly well exemplified in the case of radiation costs. As was pointed out in Chapter Two, for example, government estimates reveal that a child is three to six times more likely to contract cancer than is an adult, after both have been exposed to the same amount (one rad) of total body radiation.[25] Recognizing the impossibility of attaining complete equity in risk transference, government policymakers have recommended equity as "a *goal* to be achieved". [italics mine][26] Since federal radiation policy specifically requires that, so far as possible, inequitable risks ought to minimized,[27] it seems logical to provide for 100% nuclear insurance coverage for those bearing radiation risks. This, of course, is not the solution that has been adopted by virtue of the limit on liability. Hence the Price-Anderson treatment of this externality presupposes that principles of equity are less important than the need to protect industry against lawsuits, the low probability of a catastrophic accident, and the desirability of employing an economically 'affordable' plan.

2.2. THE HAZARDS OF LOW-LEVEL RADIATION

Like government policy concerning compensation for the risk of a catastrophic nuclear accident, public policy governing coverage of damages caused by low-level radiation is also incomplete. For this reason low-level radiation hazards also constitute a partially-compensated externality. As was pointed out in Chapter Two, nuclear industries are required to control their emissions so long as they can be justified on a 'favorable cost-benefit analysis'.[28] They must contain low-level radiation releases so long as it costs them less than $ 1000 to avoid one man-rem of exposure to the public. If it requires more than this amount to prevent an additional man-rem of exposure, then even though containment is technically possible, radiation control is not required, provided that this causes no member of the public to receive more than 0.5 rem whole-body radiation dose in any calendar year.[29] If a utility exceeds the 0.5 standard, then the government is theoretically able to compensate for this negative externality by imposing a social or economic 'tax' on the holder of the nuclear license. Clearly, however, this means of compensation is incomplete. Since the government admits that any amount of radiation is hazardous to health,[30] and that all radiation exposures are cumulative or additive,[31] its policy of failing to account, economically, for exposures of up to 0.5 rem per person per year is somewhat questionable. To understand why this is so, consider the following case.

According to government calculations, every rad of radiation causes 0.002 genetic deaths among offspring of irradiated ancestors.[32] This means, for example, that in the 30-year lifetime of a particular reactor, if 100 persons in the environs of the plant annually receive the maximum permissible annual dose of low-level radiation (0.5 rem), then this exposure alone will cause at least three of them to produce children who will die from genetic

disorders.[33] The great tragedy is that, according to current regulations, the persons who might bear such a loss could not in principle
be compensated, since the death was caused by permissible levels
of radiation (below 0.5 rem/year).[34] In other words, government
policy concerning *compensation* for this externality is not consistent with public policy regarding *assessment* of this externality.
The magnitude of this policy is clear from some recent figures
released by the EPA. According to the Environmental Protection
Agency, fatalities throughout the world that will be caused by
legally-allowable emissions, of only four of the many isotopes
(tritium, [85] Kr, [129] I, and [239] Pu) released from the nuclear fuel
cycle, will number (respectively, 1867, 4600, 63, and 24,000) a
total of 30,530 in the next 100 years.[35]

This situation of failing to compensate for certain types of externalities caused by low-level radiation, however, is not necessarily one in which government cost-benefit presuppositions must
be said to be unethical. For one thing, with so much cancer and
genetic damage induced by various chemicals and by background
and medical exposures to radiation, additional injuries (associated
with the nuclear fuel cycle) might be hard to separate from naturally occurring health consequences of radiation. If this is so,
then it would be impossible to compensate victims because they
could not be identified. The important ethical question that
needs to be asked, therefore, is not merely whether externalities
resulting from radiation exposure below 0.5 rem/year can or ought
somewhat to be compensated. The critical question is whether
society derives substantial benefits from nuclear power to justify
the creation of a political order in which, in principle, clear violations of the rights to life and to equal protection cannot be distin-
guished from the vagaries of ill health. The important problems
surrounding this externality go beyond the scope merely of nuclear technology; they extend to public policy governing all forms
of pollution claiming the lives of anonymous victims. The diffi-

culty is not only that technology causes increased cancer and genetic damage for which the public is not, and cannot be, compensated, but also that it creates a society in which it is technically impossible to determine whose rights have been violated.

An immediate question arises as a result of analyzing this policy of only partial compensation. Why does the government allow radiation standards which "still possess the potential for serious injury and/or death"?[36] The official response is twofold. First, policymakers point out that economic considerations keep industry from perfect radiation control, even though the necessary technology exists to keep such releases at or near zero.[37] Secondly, they argue that since "current . . . limits [for radiation] are of the same magnitude as background and medical doses",[38] the health effect, "attributable to manmade releases [of radiation] will . . . be negligible compared to that due to unavoidable natural causes", e.g., background radiation. In other words, given the magnitude of cancer and genetic damage 'normally' occurring, health effects of radiation releases during the nuclear fuel cycle are 'negligible'.[39]

From the preceding explanation, it is clear that one of the ethical assumptions underlying the federal radiation policy (of ignoring negative externalities resulting from less than 0.5 rem/ year exposure) is that, on the basis of economic considerations, man-made hazards may be imposed on society provided that their magnitude is less than that of naturally occurring dangers. The presupposition of the policy is that, where economic benefits are concerned, the normal or the natural is a criterion for what is morally acceptable. As William Lowrance puts it, the assumption is that "nature is perfectly benevolent".[40] Obviously, however, if *x* numbers of fires 'naturally' occur and cause *y* numbers of fatalities, then one ought not to argue, simply on the basis of the magnitude of *y*, that it would be morally permissible to induce fires, provided that their numbers were of the same or a

lesser degree of magnitude as those naturally occurring. As G.E.
Moore pointed out, "neither the 'normal' nor the 'necessary'
should be seriously supposed to be either always good or the
only good things".[41] If it were, then at any given time, the status
quo could be said to be morally acceptable, even though in reality
it might not be.[42]

If x cancers and genetic deaths annually result from natural
radiation, then provided it is economically profitable to do so,
why is one morally permitted to cause the frequency of such
deaths to increase by a number of the same magnitude as x? So
long as this question is not addressed in cost-benefit analyses of
nuclear power, then public policy built on such analyses is open
to several charges. The most damaging of these is that technology-
related externalities have not been justified on an economic basis.
Instead the alleged acceptability of a certain externality (e.g.,
increased cancer incidence) is based wholly on the undefended
assumption that the loss is 'negligible', or, as another government
analysis put it, "insignificant compared to the eight million injuries
caused annually by . . . accidents".[43]

3. The Consequences of the Failure To Compensate

Apart from their questionable ethical assumptions and the challenge they present to the recognition of basic rights, nuclear externalities are problematic on other grounds. The fact, that compensation for these spillovers is only partial, means that industry is
less likely to engage in responsible decision-making, that resources
will probably be misallocated, that doubtful assumptions regarding
economic growth will be promoted, and that energy choices will
be misrepresented. Let us examine why each of these results is
likely to follow from the existence of nuclear diseconomics.

3.1. LESS RESPONSIBLE INDUSTRY DECISION-MAKING

If a nuclear utility is required to compensate the public neither for low-level radiation exposures below 0.5 rem/year, nor for damages of a core-melt catastrophe in excess of $ 560 million, then it is questionable whether industries will be as careful with radioactive materials as they ought. In allowing nuclear utilities to be free from the duty of complete compensation to victims of accidents caused by atomic power, public policy has clearly removed the deterrent effect of liability as it affects industry decisions regarding cost-cutting measures, siting, and larger capacity. As Charles Frankel put it, a decision is responsible when the group that makes it has to answer for it to those who are directly or indirectly affected by it.[44]

Together with taxation and regulation, legal liability is one of the main ways in which economic policy can be used to control the magnitude of diseconomies. Hence when liability is limited, say economists, the extent of disamenities is sure to be greater than when it is not.[45] One interesting case in point is that of oil spills. The only way for laws to be effective in controlling this sort of environmental damage is when they require tankers to carry insurance sufficient to cover damages. This stipulation, in turn, would provide economic grounds for giant tankers to avoid enclosed waters. such as Chesapeake Bay, where damages could be greatest.[46] Hence liability laws could directly influence the nature and extent of diseconomies such as oil spills.

In response to the charge, that liability protection might encourage irresponsible decision-making by utilities, proponents of the nuclear industry have claimed that forcing the utility to go bankrupt, in the event of a disaster, would be of 'dubious social value'.[47] While it is true that such a bankruptcy would not enable the utility in question to continue to fulfill its social and legal mandate to provide power to its service area, other ethical benefits might accrue. For one thing, victims of the disaster would be able to recoup some of their losses. Moreover the negligent utility might be prevented from operating another plant in such a way as to cause a further disaster. In other words, the bankruptcy might serve to protect other potential victims from a nuclear catastrophe. As it is, the Price-Anderson Act now enables industry to make operating decisions only on the basis of economic and technological considerations, rather than also on the basis of liability constraints.

3.2. THE MISALLOCATION OF RESOURCES

Failure to provide complete compensation for nuclear externalities is not merely dangerous, however, because it encourages the nuclear industry to be less responsible than it might be, if it bore total liability for its decisions. The basic problem with the failure to compensate, either for all effects of low-level radiation or for all consequences of a nuclear catastrophe, is that such a practice masks the real cost of atomic power.

In the case of the Price-Anderson Act, the Joint Commission on Atomic Energy admitted in 1965 that the government's indemnity was provided at a cost much lower than would be assessed commercially, were such nuclear insurance available.[48] Although it is true that the 1975 amendments to the Price-Anderson Act provide for a phasing out of government subsidy, once approximately thirty more plants are built, this subsidy exists and will continue to exist for some time. Such a continuation is all the more questionable because the 1954 Atomic Energy Act has, for the last 25 years, aroused controversy over its being a 'give away' statute. In providing indemnity where private insurers will not do so, in providing it at a much lower cost than that which would be assessed commercially, and in setting a liability limit, the government allows an opportunity for misrepresentation of 'the good' and for miscalculation of the costs and benefits of nuclear energy.

It is highly likely that atomic power is only said to be a 'profitable' way of producing electricity because it is so heavily sponsored with government funds and because utilities are not required to compensate society in full for externalities of the nuclear fuel cycle. If the $ 100 billion in subsidies and the complete cost of spillovers were taken into account, it could never be calculated as inexpensive.[49] Saunders Miller, a noted economist and investment banker, has presented a detailed analysis of why nuclear fission is one of the most costly means of generating electricity. He points out that the US government has failed to take account of numerous externalities of the nuclear fuel cycle, including the risk borne as a result of the Price-Anderson Act.[50]

Because of this masking of real costs, current public policy governing compensation for radiation-related externalities can lead to the false conception that nuclear power provides unlimited, low-cost energy. 'Cheap energy' is then able to act as an opiate which dulls the moral sense of the public and provides an apparently painless incentive to consume more and more. Without this incentive, persons might be forced to consume less. If conservation does not seem economically desirable, however, then many people will not attempt to use less electricity. For this reason the 'cheap energy' conception allows consumers to live under the illusion that we need not live simpler or less wasteful lives. It allows us to refrain from asking hard questions, such as whether it is just that North America has 6% of the world's population, but consumes 30% of the planet's energy. Likewise, with no economic incentive to consume less, we have not asked seriously whether it is ethical for the US to consume two to three times as much energy, on an average, per capita, kilowatt basis, as do other developed nations in the West.[51] Thus, indirectly, failing to compensate for all the social costs of nuclear power can lead to a misallocation of resources. This misallocation comes about, in part, because energy demand is not held down by recognition of the price paid to satisfy such a demand. Instead the cost is hidden, because government has subsidized much of it, and because individual citizens have themselves paid for the effects of externalities, e.g., with higher medical bills. For both these reasons, economists admit that "an optimum allocation of resources" will not occur so long as spill-overs and subsidies are not taken into account.[52] Mishan makes this point even more strongly, for he maintains that externalities are "the most salient factor responsible for the misallocation of our natural resources".[53] This misallocation, in turn, means that "lower income groups that have suffered more than any other group from unchecked 'development' " will continue to receive less than they need to live.[54]

If current public policy continues to sanction noncompensation for nuclear externalities, then it is likely that the real price of atomic power will not be known. If these costs are not known, then it is also probable that nuclear-generated electricity will ap-pear inexpensive. But if atomic energy seems less costly than it really is, then persons might not have substantial economic incen-tive to ask troubling questions about the need for conservation or the ethics of consumption. John Rawls, proponent of one of the most important ethical theories widely discussed currently, would probably argue that present discrepancies in patterns of energy consumption are not ethical. According to him, all persons have a duty to act according to principles of justice securing the fairest distribution of goods among all people. He argues that this duty exists by introducing the notion of the 'original position'. The 'original position' is that place held by a person who is com-pletely ignorant of his natural ability, his place in society, and all other details that might give him an advantage over his fellow humans. If one imagines what principles (for the distribution of the goods of society) he might hold, were he in the original position, one will arrive at truly just principles. This means, says Rawls, that persons in the original position would reject utilitari-anism, because no rational being would agree to accept a societal theory that might sanction a loss for himself in order to bring about a greater net good for everyone.[55] Hence if one adopted

Rawls' 'thought experiment' of the 'original position' it is unlikely that he would sanction the disparity in consumption between US citizens and other peoples in the world. The reason for rejecting such a utilitarian state of affairs is that in the original position one would not know whether he would be a US citizen or not. Hence his ignorance would provide a rational basis for adopting a principle of equity.

3.3. THE PROMOTION OF DOUBTFUL ASSUMPTIONS REGARDING ECONOMIC GROWTH

Once it became financially expedient to distinguish our energy needs from our wants, more US citizens might be ready to listen to the egalitarian views of persons like Rawls. If public policy did not mask the real costs of nuclear fission and the Price-Anderson Act, then we might have economic grounds for questioning the ethic of 'more is better'. In other words, recognizing the real costs of nuclear fission might lead us to question closely-related assumptions about the value of increasing energy consumption and the value of an expanding economy. Considering these assumptions, in turn, could have beneficial social effects far beyond the scope simply of atomic power.

If we questioned the thesis that economic growth removes poverty, for example, then we might discover that the poor rarely share in the growth of real wealth and that they are "isolated from economic growth".[56] Likewise if we questioned the thesis that economic growth enriches society, then we might find that the bad effects of growth outweigh the good. Alternatively, we might evaluate the plausibility of "the relative income hypothesis", and we might consider whether happiness is enhanced more by an expanding economy or more by improving one's relative place in it.[57] Finally, if we questioned the thesis that economic growth checks inflation, then we might discover that inflation is greater in faster growing economies than it is in slower ones.[58] As it is, however, our policies governing externalities have neither forced these troublesome queries upon us, nor delivered us from the ethical and economic problems that ignoring such questions can bring.

Accurate evaluations of nuclear externalities might also force Americans to ask some critical questions which are political, as well as economic and ethical. Is it morally desirable to ignore billions of dollars of spillover effects and subsidies, and thereby encourage development of a public resource, nuclear power, which has been turned over to the employment of a monopolistic segment of private industry?[59] Why have public funds continued to be poured directly and indirectly into private enterprise, despite the fact that the public has little real control over nuclear power? One reason might be that the public has not analyzed all the social and political costs of pursuing accelerating energy consumption. The electric power industry already enjoys antitrust immunity and the elimination of competition. If these benefits are of questionable ethical value, then so are nuclear externalities. Indirectly they increase the utility's power, which from the point of view of the consumer, means that it is less likely that the industry can ever be challenged in its behavior or policies.

3.4. THE MISREPRESENTATION OF ENERGY CHOICES

Besides helping to create a false economic security within which important ethical questions (related to conservation and control of public policy) do not naturally arise, nuclear externalities also contribute to other undesirable consequences. For one thing, they preclude a fair assessment of the merits of one energy technology with respect to another. So much money has been poured into nuclear fission, directly and indirectly, that the safer, more environmentally sound technologies, such as solar power, have taken a second seat because of lack of funding. As a result, the most ethical, equitable and justifiable energy technologies might not be chosen to supply our energy.[60] For example, the US might be 'locked into' a breeder technology, not because it is desirable on grounds of equity and safety, but because it is needed to provide fuel for the light water reactors. Obtaining fuel might be seen as necessary, if it is the only way to get a 'return' on enormous capital investments (subsidies) in fission technology.

What the above argument comes down to is that the failure to compensate for externalities of the nuclear fuel cycle provides an ethical context in which it is increasingly tempting to 'sell our souls' in a Faustian bargain for cheap and abundant energy. What is wrong is not only that in such a tradeoff we ignore questions related to the ethics of consumption and the ethics of leaving a legacy of radioactive waste for generations to come. What is also wrong is that, like every Faustian bargain, what we get is not really cheap. We receive only the illusion of inexpensive power because the public and the government have indirectly picked up the tab. Such misinterpretations are dangerous, and if nuclear externalities lead to them, they are dangerous. The long-term consequences of these illusions are that it is more likely that Huxley's *Brave New World* will really come to pass. Hence we might, because of a false economic security, fail to evaluate all our energy choices. And if we fail in this, we may be forced inadvertently to sacrifice individual rights and freedoms for the energy promised by nuclear fission.

4. The Consequences of Recognizing Amenity Rights

On the other hand, if we begin now to compensate for nuclear externalities by such means as removing the Price-Anderson liability limit, then we will have a more equitable and accurate picture of all the energy options available to us. Changing public policy in this manner, so as to provide complete compensation for nuclear externalities, might require the recognition of a class of amenity rights, rights of a person not to "be forced against his will to absorb . . . noxious by-products of the activity of others".[61]

While the particulars of such a system of amenity rights would be difficult to work out, it is clear that recognition of them is desirable for both ethical and economic reasons. Let us examine some of the consequences that would follow from recognition of these rights.

For one thing, citizens might be prohibited from causing any spillovers for which they were unable to compensate others. If a person using a loud radio or power mower were unable to compensate his neighbors who enjoyed being outside, for example, then he might be unable to use his radio or lawn mower.[62] While his freedom might be restricted thus by the amenity rights of others, their freedom to enjoy their property might be enhanced. As Mishan argues, such a situation would be highly desirable. For too long, he says, citizens have been 'robbed of choice' regarding such things as quiet and attractive surroundings. For too long, he says, the public has been accepting the physical environment as it accepts the weather. This situation must change, notes Mishan, and law must enable us to compensate for externalities; otherwise, markets will not flourish in an orderly fashion, and the real costs of goods will be misrepresented.[63]

Another consequence of recognizing amenity rights is that private enterprise will be less likely to ignore the welfare of the public. If externalities are compensated and amenity rights considered, then industry will have an incentive for attempting to protect society from the ill effects of its actions. Of course one problem with this consequence is that an ethical interpretation of what constitutes "ignoring the welfare of the public" will be difficult to obtain, particularly because many interpretations will be challenged as infringements of property rights. As Mishan points out, private property has been regarded as inviolate for centuries; if public policy requires property owners and industrialists to take account of disamenities caused by their use of their property, then a whole new theory of property rights will be required to justify the new policy.[64]

A further consequence of implementing a system of amenity rights is that the rich and poor alike will have more equal access to desirable environmental quality. Without recognition of these rights, only the rich are able to afford to avoid certain spillover effects. Only the rich can decide to move, for example, when a nuclear plant is built nearby. As Mishan points out, "the richer a man is the wider is his choice of neighborhood". But if amenity rights were recognized, then living in an environmentally desirable neighborhood would not be largely a function of one's ability to pay for enough land, in order to insulate himself from many noxious externalities. "Thus the recognition of amenity rights would have favourable distributive effects on the welfare of society"; it would enhance the environment generally but would help most the poor who have borne the worst effects of faulty distribution of resources.[65]

5. Conclusion

Although recognition of amenity rights would lead to a number of desirable consequences and would help take account of the undesirable effects of externalities, this alone might not solve the problems created by the externalities of the nuclear fuel cycle. If my remarks about the risks created by low-level radiation and by the possibility of core-melt catastrophe are correct, then the dilemma posed by nuclear generation of electricity cannot be resolved simply by compensating for the social costs incurred, although compensation is a first step toward resolving it. The deeper problem is that the very nature of radiation-related externalities, apart from whether they are accounted for within cost-benefit analyses, precludes certain identification of radiation-caused injuries. For this reason these externalities challenge society's ability both to identify the victims of injustice and to recognize the difference between equity and utility. This means that, if we continue our current economic policies governing nuclear technology, then we may abandon not only the practical ability to detect violations of basic rights, but also our recognition of those rights.

Notes

[1] Arthur Okun, *Equality and Efficiency: The Big Tradeoff*, The Brookings Institution, Washington, D.C., 1975.

[2] D.C. North, 'Political Economy and Environmental Policies', *Environmental Law* 7(3), (Spring 1977), 449.

[3] The classic example of a negative externality was given by the Cambridge economist, Arthur Cecil Pigou, who spoke of the factory chimney which spews soot on surrounding houses. (*The Economics of Welfare*, Macmillan, London, 1962.) For further discussion of environment-related externalities, see E.B. Mishan, 'Property Rights and Amenity Rights', in E.B. Mishan, *Technology and Growth: The Price We Pay*, Praeger, New York, 1970, pp. 37ff.; hereafter cited as 'Property Rights'. See also Richard B. Stewart, 'Paradoxes of Liberty, Integrity, and Fraternity: The Collective Nature of Environmental Quality and Judicial Review of Administrative Action', *Environmental Law* 7(3), (Spring 1977), 463–484 (hereafter cited as: 'Paradoxes'); R.T. Roelofs, J.N. Crowley, and D.L. Hardesty, 'Environment and the New Economics', in *Environment and Society* (ed. by Roelofs *et al.*), Prentice-Hall, Englewood Cliffs, 1974, pp. 113–18; and B. Commoner, *The Closing Circle*, Bantam, New York, 1974, pp. 249–91.

[4] Mishan, 'Property Rights', p. 37.

[5] Stewart, 'Paradoxes', p. 463.

[6] Mishan, 'Property Rights', p. 42.

[7] This point is discussed in Stewart, 'Paradoxes', pp. 464–84; and in E.J. Mishan, *Cost-Benefit Analysis*, Praeger, New York, 1976, pp. 133–38; hereafter cited as: *Cost*.

[8] See note 1.

[9] See, for example, Ivan Illich, *Energy and Equity*, Harper and Row, New York, 1974.

[10] See, for example, K.L. Pavitt, 'Malthus and Other Economists', in *Thinking about the Future: A Critique of 'The Limits to Growth'* (ed. by H.S.D. Cole *et al.*), Chatto and Windus, London, 1973, pp. 137–58.

[11] See, for example, D. Meadows *et al.*, 'Pollution and the Limits to Economic Growth', *The Limits to Growth*, New American Library, New York, 1974, pp. 74–94, 137–51.

[12] See, for example, John Maddox, *The Doomsday Syndrome*, Macmillan, London, 1972, pp. 186–214.

[13] As was pointed out in Chapter One, Section 3, the nuclear fuel cycle includes all processes necessary to the production of electricity by nuclear fission. It involves mining uranium ore, milling the uranium out of the ore, refining it to purity standards, enriching the uranium by means of a partial separation of the isotopes, converting the uranium from hexafluoride, fabricating it into fuel elements, possible reprocessing of the spent fuel, and storing the radioactive wastes. For further information, see J.F. Hogerton, *Atomic Fuel*, US Atomic Energy Commission, Washington, D.C., 1964.

[14] US Atomic Energy Commission, *Nuclear Power and the Environment*, US Government Printing Office, Washington, D.C., 1969, p. 28. This same policy is found throughout both government literature and nuclear industry material. See also, for example, W.O. Doub, 'Meeting the Challenge to Nuclear Energy Head-On', *Atomic Energy Law Journal* 15(4), (Winter 1974), 238–64; and Mason Willrich, *Global Politics of Nuclear Energy*, Praeger, New York, 1971, p. 73.

[15] US AEC, *Nuclear Fuel Resources and Requirements*, WASH-1243, T1D UC–51, US Government Printing Office, Washington, D.C., 1973, p. 45. See also US AEC, *Nuclear Fuel Supply*, WASH-1242, UC–51, US Government Printing Office, Washington, D.C., 1973, p. 1.

[16] *Economic and Environmental Implications of a US Nuclear Moratorium*, 3 Vols., Institute for Energy Analysis, Oak Ridge Associated Universities, Oak Ridge, 1976; see also A.L. Hammond, 'Nuclear Moratorium: Study Claims That Effects Would Be Modest, Foresees Low Growth Rate for Total Energy Demand', *Science* 195 (4274), (January 1977), 156–57.

[17] For capacity statistics, see D.D. Comey, 'Will Idle Capacity Kill Nuclear Power?', in *Countdown to a Nuclear Moratorium* (ed. by Richard Munson), Environmental Action Foundation, Washington, D.C., 1976, pp. 65–79; and V. McKim, 'Social and Environmental Values in Power Plant Licensing', *Values in the Electric Power Industry* (ed. by K.M. Sayre), University Press, Notre Dame, 1977, pp. 217–18. For data on uranium prices, see P. L. Joskow, 'Commercial Impossibility, The Uranium Market, and the Westinghouse Case', *Journal of Legal Studies* 6 (1), (January 1977), 119–76. For information on rising construction costs and on causes of cancellation, also see Joskow; R. Nader, 'Nuclear Power Collapsing', *Critical Mass Journal* 4 (4), (July 1978), 18–19; and J.J. Berger, *Nuclear Power*, Laurel, New York, 1976, pp. 112ff.

[18] 'Utilitarianism and Cost-Benefit Analysis: An Essay on the Relevance

of Moral Philosophy to Bureaucratic Theory', in *Values in the Electric Power Industry* (ed. by K. Sayre), University Press, Notre Dame, 1977, p. 217.
[19] 'Theoretical Possibilities and Consequences of Major Accidents in Large Nuclear Power Plants', US AEC Report WASH-740, Government Printing Office, Washington D.C., 1957; and R.J. Mulvihill, D.R. Arnold, C.E. Bloomquist, and B. Epstein, 'Analysis of United States Power Reactor Accident Probability', WASH-740 update, PRC R-695, Planning Research Corporation, Los Angeles, 1965. Cf. US AEC, Summary Report, 'Reactor Safety Study: An Assessment of Accident Risks in US Commercial Nuclear Power Plants [Rasmussen Report, WASH-1400]', *Atomic Energy Law Journal* **16** (3), (Fall 1974), 201–202; hereafter cited as: 'WASH-1400-Summary'. Since nuclear plants today are approximately five times the size of those described in the Brookhaven Report (WASH-740), it is possible that accident consequences could be more severe than those cited from WASH-740. See Berger, *Nuclear Power*, p. 45.
[20] 'AEC Staff Study of the Price-Anderson Act, Part I', *Atomic Energy Law Journal* **16** (3), (Fall 1974), 220. For further information on the total insurance now available, see Chapter Four, Section 1.   ·
[21] This fact is verified by J. Marrone, 'The Price-Anderson Act: The Insurance Industry's View', *Forum* **12** (2), (Winter 1977), 607; W.S. Caldwell *et al.*, 'The "Extraordinary Nuclear Occurrence" Threshold and Uncompensated Injury Under the Price-Anderson Act', *Rutgers-Camden Law Journal* **6** (2), (Fall 1974), 379; R. Lowenstein, 'The Price-Anderson Act: An Imaginative Approach to Public Liability Concerns', *Forum* **12** (2), (Winter 1977), 596; A.W. Murphy and D.B. LaPierre, 'Nuclear "Moratorium" Legislation in the States and the Supremacy Clause', *Environment Law Review 1977* (ed. by H.F. Sherrod), Clark Boardman, New York, 1977, p. 405; and 'AEC Staff Study of the Price-Anderson Act', p. 209.
[22] According to the Rasmussen Report (US AEC, *Reactor Safety Study: An Assessment of Accident Risks in US Commercial Nuclear Power Plants*, WASH-1400, Government Printing Office, Washington, D.C., 1974, p. 157, the per-year, per-reactor probability for a core melt is 1/17,000. Using this figure, one can compute the probability that a core melt will occur during the 30-year lifetime of one of the 65 plants now operating or one of the 70 plants now under construction. Employing the formula for the probability of mathematically independent events, P (a core melt in at least one of the 135 plants over a 30-year lifetime) = $1 - P$ (no core melt in any of the 135 reactors over a 30-year lifetime), one obtains P (core melt) = $1 - (1 - (1/17,000))^{4050} = 1 - (0.99994)^{4050} = 1 - 0.78427 = 0.2157$. Thus there is a 20% probability of a core melt in one of 135 reactors now operating or under construction.
[23] Michael Shannon (Environmental Protection Agency), 'The Dilemma of

Liability and Perpetual Care Issues', US EPA, *Waste Management Technology and Resource and Energy Recovery*, Government Printing Office, Washington, D.C., 1977, p. 353.

[24] 'Ethics and Population Limitation', in *Philosophical Problems of Science and Technology* (ed. by A.M. Michalos), Allyn and Bacon, Boston, 1974, p. 560. See Chapter Two, Sections 3.1 and 3.2, for further discussion of utility and equity.

[25] US AEC, *Comparative Risk-Cost-Benefit Study of Alternative Sources of Electrical Energy*, WASH-1224, Government Printing Office, Washington, D.C., 1974, p. 4-14; hereafter cited as:'WASH-1224'. One 'rad' is a measure of the dose of any ionizing radiation to body tissues in terms of the energy absorbed per unit mass of the tissue. See Chapter Two, Section 3.1 and 3.2., for further discussion of this example.

[26] US Environmental Protection Agency, *Considerations of Environmental Protection Criteria for Radioactive Waste*, Government Printing Office, Washington, D.C., 1978, p. 48; hereafter cited as *Considerations*.

[27] S. Lichtman (US EPA, Waste Environmental Standards Program), 'Risk in Radioactive Waste Management', in US EPA, *Proceedings of a Public Forum on Environmental Protection Criteria for Radioactive Wastes*, ORP/CSD-78-2, Government Printing Office, Washington, D.C., 1978, pp. 34ff. Environmental Protection Agency, 'Criteria for Radioactive Wastes', *Federal Register* 43 (221), (November 1978), 53266–53267.

[28] Nuclear Regulatory Commission, *Issuances*, Government Printing Office, Washington, D.C., 1977, 5, Book 2, p. 928; hereafter cited as: *Issuances*.

[29] *Code of Federal Regulations*, Government Printing Office, Washington, D.C., 1978, 10, Part 20, p. 189. NRC, *Issuances*, p. 980. According to '10 CFR 20', p. 184, the *rem* "is a measure of the dose of any ionizing radiation to body tissues in terms of its estimated biological effect relative to a dose of one roentgen of X-rays"; one *man-rem* of radiation is equivalent to one person's exposure to one rem of radiation; two man-rems are equivalent to one person's exposure to two rems or to two persons' exposures to one rem, and so on. For most purposes, a rem is equivalent to a rad. A radiation *dose* is the quantity of radiation absorbed, per unit of mass, by the *whole body* or by any portion of the body. (See note 25 for the definition of a rad.)

[30] US AEC, 'WASH-1224'. This is also the position of the International Commission on Radiation Protection (on which federal regulations are based), the US Public Health Service, and the Federal Radiation Council; see US AEC, *Nuclear Power and the Environment*, Government Printing Office, Washington, D.C., 1969, p. 41; hereafter cited as: AEC, *Nuclear Power*. See EPA, *Considerations*, p. 30, and Chapter Two, Section 2.

[31] 'WASH-1224', p. 4-13; O. Hansen, 'Development and Application of Radiation Protection Standards', *Idaho International Review* 12 (1), (Fall

1975), 1–32; M.S. Young, 'A Survey of the Governmental Regulation of Nuclear Power Generation', *Marquette Law Review* **59** (4), (1976), 842–43.

[32] 'WASH-1224', p. 4-14.

[33] Using the statistics from 'WASH-1224', one calculates: (0.002 genetic deaths/rem) × (30 × 0.5 rem) = 3% risk per person, or 3 deaths per hundred exposures. Such a prediction is possible because of the linear dose-response effects of radiation exposure. That is, a certain quantity of response (cancer, genetic damage, etc.) will result per unit (man-rem) of population dose. For example, "the same total radiation effect would be predicted for a group of one million people receiving 0.1 rem per person in one hour as for a group of one hundred million people receiving 0.001 rem per person in one year". ('WASH-1224', p. 4-13.) See also Chapter Two, Section 3.2.

[34] The discussion here focuses on 'in principle' compensation, since as will be seen later, it might not be possible to identify whether an injury were the result of background radiation or whether it were the consequence of nuclear plant releases. Hence compensation 'in fact' might be impossible.

[35] Energy Research and Development Administration, *Final Environmental Impact Statement: Waste Management Operations, Idaho National Engineering Laboratory*, ERDA-1536, National Technical Information Service, Springfield, Virginia, pp. III-103 and III-104.

[36] James Elder, 'Nuclear Torts . . . And the Potential for Uncompensated Injury', *New England Law Review* **11** (1), (Fall 1975), 113.

[37] AEC, *Nuclear Power*, pp. 42, 48; US EPA, *Considerations*, p. 14.

[38] 'WASH-1224', p. 4-8.

[39] 'WASH-1224', p. 4-7. See Chapter Six, especially Sections 2.31, 2.32, and 2.33 for discussion of 'negligible hazards' of nuclear power.

[40] *Of Acceptable Risk*, William Kaufmann, Los Altos, 1976, p. 85.

[41] *Principia Ethica*, University Press, Cambridge, 1951, pp. 58, 43. For further discussion of this point, see Chapter Six, especially Sections 2.12 and 3.33.

[42] See Chapter Two, Section 3.3.

[43] 'WASH-1400', pp. 182–83.

[44] Cited by G. Hardin, 'To Trouble a Star: The Cost of Intervention in Nature', in *Environment and Society* (ed. by R.T. Roelofs, J.N. Crowley, and D.L. Hardesty), Prentice-Hall, Englewood Cliffs, N.J., 1974, p. 125.

[45] See Mishan, *Cost*, pp. 139–44; V. Morgan and A. Morgan, *The Economics of Public Policy*, University Press, Edinburgh, 1972, p. 64; hereafter cited as: *Economics*.

[46] This example is given by D.C. North and R.L. Miller, *The Economics of Public Issues*, Harper and Row, New York, 1971, pp. 75–79; hereafter cited as: *Issues*.

47 F.H. Schmidt and D. Bodansky, *The Energy Controversy*, Albion, San Francisco, 1976, p. 146.
48 'AEC Staff Study of the Price-Anderson Act, I', p. 239.
49 Berger, *Nuclear Power*, pp. 94–97, 106–112, 144–47; see also J. Gofman and A. Tamplin, *Poisoned Power*, Rodale, Emmaus, 1971, pp. 177, 199; J. Primack and F. Von Hippel, 'Nuclear Reactor Safety', *The Bulletin of the Atomic Scientists* **30** (8), (October 1974), 7; E. Muchnicki, 'The Proper Role of the Public in Nuclear Power Plant Licensing Decisions', *Atomic Energy Law Journal* **15** (2), (Spring 1973), 45. Berger, p. 112 explains that "more than a hundred billion dollars [in federal funds] has already been invested in nuclear power" subsidies since 1945.
50 Saunders Miller, *The Economics of Nuclear and Coal Power*, Praeger, New York, 1976, p. 105.
51 These statistics are taken from M. Mesarovic and E. Pestel, *Mankind at the Turning Point: The Second Report to the Club of Rome*, Signet, New York, 1974, pp. 135–139.
52 Morgan and Morgan, *Economics*, pp. 63–64.
53 E.J. Mishan, *The Costs of Economic Growth*, Praeger, New York, 1967, p. 55; hereafter cited as: *CEG.*
54 E.J. Mishan, *Technology and Growth: The Price We Pay*, Praeger, New York, 1969, 41; hereafter cited as: *Technology.*
55 Rawls, *A Theory of Justice*, Harvard University Press, Cambridge, 1971. For further discussion of Rawls' view, see Chapter Two, Sections 3.1 and 3.2.
56 E. J. Mishan, *21 Popular Economic Fallacies*, Praeger, New York, 1969, p. 236; hereafter cited as: *Fallacies.*
57 Mishan, *Fallacies*, p. 245.
58 Mishan, *Fallacies*, pp. 222–30.
59 J. Palfrey, 'Energy and the Environment', *Columbia Law Review* **74** (8), (December 1974), 1391.
60 M. Willrich, *Global Politics of Nuclear Energy*, Praeger, New York, 1971, p. 181, uses this argument regarding nuclear fusion. S. Novick, *The Electric War*, Sierra, San Francisco, 1976, p. 318, makes the same point in terms of solar energy.
61 Mishan, *Technology*, p. 39.
62 This point is made by Mishan, *Technology*, p. 38; and Mishan, *CEG*, p. 55.
63 Mishan, *Cost*, pp. 128–129.
64 Mishan, *Technology*, p. 37.
65 Mishan, *Technology*, p. 41.

Chapter Six

Nuclear Safety and the Naturalistic Fallacy

The practical world of industry, technology, and government is haunted by numerous theoretical errors. One of the most ominous of these, 'the naturalistic fallacy', informs and guides the actions of persons who take themselves to be hardheaded, pragmatic, free of theory, and common-sensical in making public policy regarding technology. Since this error is usually unrecognized, it is particularly dangerous because it often goes uncriticized.[1] My object in this chapter is to point out, analyze, and therefore exorcize the methodology responsible for a peculiar type of moral blindness known as the 'naturalistic fallacy'. As we shall see, this fallacy plays a key role in the technology assessments which are used to support current public policy governing nuclear power. Perhaps more than any other methodological problem, the naturalistic fallacy is responsible for the ethically questionable assumptions criticized thus far in the book. Although part of this chapter requires the reader to follow the particulars of somewhat abstract logical analysis, it is of great importance in understanding the precise problems inherent in assessment both of nuclear technology and of many other technologies where this same difficulty arises.

1. The Naturalistic Fallacy

'The naturalistic fallacy' is the name given to a particular error

often committed when one uses inappropriate evidence to sub-
stantiate an ethical or policy conclusion. Although it is impossible,
as G.E. Moore noted, to attain certainty in ethics, "nevertheless
the *kind* of evidence, which is both necessary and alone relevant"
to the proof or disproof of any ethical proposition, "is capable of
exact definition".[2] Moore was convinced that the sciences alone
could never provide such evidence, since there was a "distinct class
of ethical judgments", not deducible from nonethical premisses.[3]
According to him, anyone, who attempted to draw ethical conclusions solely from empirical observation and induction, was
guilty of the naturalistic fallacy.[4]

In general both Moore and Frankena (author of the most famous post-Moorean analysis of the fallacy) agree that the naturalistic fallacy rests upon the failure to realize that ethical characteristics and propositions are different in kind from nonethical ones.[5]
Hence the naturalistic fallacy is committed whenever one attempts
to define ethical characteristics in nonethical terms or to deduce
ethical propositions from nonethical ones. Those who commit this
fallacy do so because they specifically assume, wrongly, that ethics
is no more than "an empirical or positive science".[6]

1.1. THREE SPECIES OF ERROR

Considered as the attempt to reduce ethical characteristics or
propositions to nonethical ones, the naturalistic fallacy may take
any one of a number of forms. Chief among these are (1) replacing
ethics with one of the natural sciences;[7] (2) deriving "ought (evaluative, normative, emotive, or prescriptive)" statements from "is
(nonevaluative, descriptive, or factual)" statements;[8] and (3)
failing to consider the 'open question'.[9]

Error (1), attempting to replace ethics with one of the natural
sciences, is committed whenever one attempts to give scientific
reasons, alone, as a justification for ethical beliefs. If one argued,
for example, that taking risk *A* is moral, solely because risk *A* has

a low probability of causing catastrophe, then one would commit this error. He would be assuming that a purely scientific
property, i.e., a low probability, constituted a sufficient condition for terming a risk 'acceptable'. Such a procedure is erroneous,
because it ignores the differences between science and philosophy.
As Moore points out, "ethical judgments about the effects of
action involve a difficulty and a complication far greater than
that involved in the establishment of scientific laws". Ethical
questions are such that "a correct answer to any of them involves
both judgments of what is good in itself and causal judgments".[10]
Hence, for Moore, empirical or inductive considerations represent only a part of what must be addressed in making ethical judgments.

Error (2), attempting to derive 'ought' statements from 'is'
statements, is said by many to be wrong because it allows one to
derive normative or prescriptive propositions from descriptive or
factual ones. According to most ethicians, this is impossible
because no amount of factual or nonethical information, alone,
constitutes sufficient grounds for an ethical conclusion. As Moore
points out, information that something is (descriptively) 'desired',
is not alone grounds for saying it is (normatively) 'desirable'.[11]
Or, for example, if it were the case that a majority of people
committed action *A*, this fact would not be reason enough for
asserting that one ought to commit *A*. According to Moore, it is
wrong to say "that we ought to . . . [do something] (an ethical
proposition), because we actually do . . . it."[12] Hence to maintain,
that descriptive or 'is' propositions alone provide adequate bases
for normative or 'ought' conclusions, is to commit the naturalistic
fallacy.[13]

Error (3), failing to consider 'the open question', has been
considered by authors such as Kohlberg to be another variant of
the naturalistic fallacy.[14] ('The open question' is essentially the
argument that, no matter what natural quality of a thing is defined

as good, "it is always an open question whether or not the quality" is *in fact* good.[15] In other words, whenever something is *defined* as good, one can always ask for reasons why this is so, or challenge the assumptions upon which the definition rests.) One ignores 'the open question' whenever one *defines* something as good simply because it has some natural property. For example, suppose I define something as good because it has the natural property of causing the G.N.P. to rise. My definition is said to be 'fallacious' because, by virtue of the fact that it is a *definition*, it ought not to be open to challenge. Yet it is open to question. One is always able to ask whether "whatever causes the G.N.P. to rise" is good. Hence it is said to be wrong to ignore the openness of alleged definitions of what is good.[16] When one does so, he commits one variant of the naturalistic fallacy.

1.2. THE SIGNIFICANCE OF THE FALLACY

Although there has been some controversy regarding the view that commission of any of these three variants of the naturalistic fallacy is a species of error, even opponents of this belief have admitted that the naturalistic fallacy has 'remained alive', since Moore introduced it in 1903 in his *Principia Ethica.*[17] "The majority of philosophers still view the fallacy as a yawning chasm."[18]

There are several reasons why a majority of philosophers have continued to view the naturalistic fallacy as an error. As was mentioned earlier, commission of the fallacy precludes providing sufficient reasons for ethical decisions.[19] Also, in attempting to replace ethics with science, it "is inconsistent with the possibility of any ethics whatsoever".[20] Moreover, in defining ethics in terms of only natural or empirical phenomena, one who commits this fallacy locks himself into a rigid dogmatism that limits the openness and completeness desirable in ethical discussion.[21]

2. Commissions of the Fallacy in Government Studies of Nuclear Power

Although the nature of this fallacy bears much reflection, I shall not spend further time discussing it here. My purpose is not to engage in theoretical ethical analysis but to investigate its applications to technology assessment and public policy-making. Specifically, I intend to show how commission of this classic fallacy in ethics has undermined authentic technology assessment regarding nuclear power. My point, however, is not simply to argue for or against current policy governing nuclear technology, but to explain why this particular type of methodological error ought to be avoided. First I will attempt to explain what implausible methodological assumptions are built into each instance of the alleged fallacy; secondly, I will discuss the undesirable ethical consequences that follow from these commissions; and thirdly, I will outline a desirable philosophical procedure for avoiding this fallacy in technology assessments and public policy evaluations in the future.

On the grounds that they are the basis for public policy decisions, several assessments of nuclear technology are of overarching importance. The most significant of these is the only allegedly complete government study of nuclear reactor safety, the Rasmussen Report, known as WASH-1400. As was explained in Chapter Four, it is the basis for acceptance of all current light-water, fission technology. Another study of nuclear energy is that by Hans Bethe, "The Necessity of Fission Power".[22] Although numerous scientists have done technical studies in this area, I have selected the Bethe analysis because its author, a Nobel Prize winner, has outstanding credentials in the area of physics.

Bethe examines a number of questions concerning the desirability of employing nuclear fission to generate electricity. The bulk of his remarks is addressed to seven questions, the second of which is not only the key one addressed by Bethe but also the major topic of inquiry in the Rasmussen Report. It is central, both to the single most important assessments of nuclear technology and to continuing public debate over health and safety. This question is whether we ought to accept the risk of a major reactor accident releasing catastrophic amounts of radioactivity.[23] In answering this question affirmatively, Bethe and the authors of WASH-1400 employ three major arguments:

(2.1)   that, since the core melt probability is approximately 5×10^{-5} per reactor per year, the public has only a small risk from a possible nuclear accident;[24]

(2.2)   that, since society accepts other risks whose consequences are more disastrous than those of nuclear power, the risk of a major nuclear accident also ought to be accepted;[25] and

(2.3)   that, although a nuclear accident would cause 5000 cancer deaths, 3000 cases of genetic damage, and numerous instances of thyroid disease, since "the small increases in these diseases would not be detected" and would be "insignificant compared to the eight million injuries caused annually by other accidents," the risk of a nuclear accident ought to be accepted.[26]

2.1. THE ARGUMENT BASED ON PROBABILITY OF A CORE MELT

The first argument above many be formulated in the following manner:

(2.1.1) All risks of major reactor accidents are risks of human-caused, catastrophic consequences (of otherwise beneficial technology) which have a very low probability of occurring.

(2.1.2) All risks of human-caused, catastrophic consequences

(of otherwise beneficial technology), which have a very
low probability of occurring, are risks which ought to be
accepted.

(2.1.3) All risks of major reactor accidents are risks which ought
to be accepted.[27]

As stated by Rasmussen and Bethe, this argument does not contain
premise (2.1.2); it is of the form "All *A* is *B*, therefore all *A* is *C*".
Since any argument of this type, taken strictly, is invalid, they
must have intended to present an enthymeme containing a suppressed premise (2.1.2).[28] With (2.1.2), the argument is clearly
logically valid. If there is a naturalistic fallacy involved here, it has
nothing to do with a purely formal error.

(2.1.1) is a statement whose truth may be determined solely on
the basis of mathematics and science. Although Bethe, following
Rasmussen, supports this statement by the estimate that a meltdown has a probability of one in 17,000, there are several reasons
to doubt this figure,[29] including the fact that it has been criticized
by the American Physical Society and the Environmental Protection Agency.[30] The correctness of (2.1.1), however, has no
bearing on whether the naturalistic fallacy is committed in argument (2.1). Therefore, for purposes of argument, let us assume
that (2.1.1) is both true and the product of the best available
mathematical and scientific methodology. This leaves us with
(2.1.2), the premise suppressed by both Rasmussen and Bethe.

2.1.1. *Low Probability of Catastrophe as a Sufficient Condition for Acceptable Risks*

Since no arguments are given to support (2.1.2), both authors seem
to have intended it to be accepted as true by definition. According
to G.E. Moore, however, employing such a premise constitutes
commission of the naturalistic fallacy. This is because Rasmussen
and Bethe have named *properties* (viz., being beneficial to the

economy and having a low probability of causing a health risk) of a *good* thing (an increasing rate of power generation), and have equated these mathematical and scientific properties with what is good. Such an equation can be shown to be erroneous since, in general, it is incorrect to argue that (2.1.2) is the case. For example, it would be ethically reprehensible to subject all persons to chest X-rays on a routine basis every month, in order to guard against tuberculosis. Although this procedure would have a low probability of causing a catastrophic number of cancer deaths, such a radiation risk is obviously not one that ought to be taken by all persons. Clearly some actions, possessing such a low probability, are good, while others are not worth the risk. Since not all consequences (of an otherwise beneficial technology), having a low probability of being catastrophic, may be said to be good or to be things which ought to be accepted, thesis (2.1.2) cannot be assumed, in general, to be true.

In *defining* several effects of technology (economic benefit and low health risk) as acceptable, Bethe and Rasmussen have ignored the question of how and why the assets and liabilities of nuclear power ought to be balanced in a particular way. If one attempts to argue cogently that one ought to take risk X, then one cannot merely compute risk X and then define this result as agreeable because of economic benefits that result. Instead one must argue why this risk, or any other, is worthy of assent. In failing to realize that "risk X has a very low probability of resulting in catastrophe and a very high probability of inducing economic benefits" is not equivalent to "risk X ought to be accepted", Bethe and Rasmussen fail to tell us precisely *why* one ought to consent to taking this risk. In so doing they assume, rather than prove, that they have found a sufficient condition for approving the nuclear risk. One of the difficulties inherent in their methodological error, begging the question, is that certain relevant considerations are not treated. They ignore whether the same group of people who bear the risk of a core melt will be those who benefit from the electricity generated by nuclear energy. Likewise, in response to their 'definition' of acceptable risk, one could merely question: "But is the risk moral, since a nuclear accident victim might be limited, by the Price-Anderson Act, to collecting only 3% of his property losses?"[31] "Is the risk acceptable, since persons within the environs of a nuclear plant could be deprived of Fifth and Fourteenth Amendment guarantees of protection of property?"[32] Although asking such questions might not have changed their conclusion (2.1.3), Bethe and Rasmussen were clearly wrong in defining the problem, *a priori*, as not involving questions of equity, property rights, and due process. While not desirable, however, their omission is understandable. Bethe and Rasmussen are scientists engaged in a scientific assessment, not ethicians involved in an ethical analysis. While they need not have addressed nonscientific parameters in their studies, nevertheless they might have been consistent and therefore avoided drawing ethical conclusions without having the benefit of ethical analyses.

2.1.2. *What Is Normal as a Criterion for What Is Moral*

I suspect that if Bethe and Rasmussen were to respond to these charges, they would say that nuclear technology is necessary. After all, Bethe's article is entitled, "The Necessity of Nuclear Fission". Consequently they might argue that desperate problems require desperate solutions, and that because of the energy crisis, the nuclear risk ought to be taken. Although neither Bethe nor Rasmussen provides such an analysis of why nuclear power is necessary, Bethe does say: "This country needs power to keep its economy going".[33] However, "neither the 'normal' nor the 'necessary' could be seriously supposed to be either always good or the only good things".[34]

Obviously *A* may be necessary for *B*, but unless we know why *B* is good, then there is no reason to believe that we ought to seek

A. Both Bethe and Rasmussen ignore the question of why *B* (maintaining the current levels of energy usage) is good when they argue that *A* (nuclear energy) is necessary in order to obtain *B*. Noted economists Ezra Mishan and E.F. Schumacher believe that, in this case, *B* is not good. Hence Bethe and Rasmussen appear to have made a glaring omission. Whether the present state of the economy (with the US using, misusing, and wasting two to three times the per capita amount of energy as other Western developed nations) is good,[35] is an open question. Since Bethe and Rasmussen have assumed, rather than argued, that the current level of energy usage is good, and that nuclear power is necessary to maintain this level, they have again employed the naturalistic fallacy in giving a definition of what is good rather than reasons why it is so.

2.2. THE ARGUMENT BASED ON OTHER PROBABLE ACCIDENT RISKS

Their second argument, (2.2) that nuclear accident risks are low compared to other risks having beneficial effects, and hence that nuclear technology ought to be accepted, fares no better than the first. Like (2.1), this argument, if it is not logically fallacious, must rest on an enthymeme containing an explicit premise, a suppressed premise and a conclusion:

(2.2.1) All risks of death or injury from nuclear accidents are risks which are lower than other human-caused accident risks already accepted by society.

(2.2.2) All risks which are lower than other human-caused accident risks already accepted by society are morally acceptable.

(2.2.3) All risks of death or injury from nuclear accidents are morally acceptable.[36]

Although the explicit premise (2.2.1) is of doubtful truth (see note 25), let us assume (for purposes of argument) that it is true, since its veracity is a matter of mathematical and statistical accuracy, discussion of which is not within the main purpose of this chapter. From the point of view of ethical methodology, however, the suppressed premise, (2.2.2), is very interesting.

2.2.1. *Consistency as a Sufficient Condition for Acceptable Judgments*

If consistency were both a necessary and a sufficient criterion for doing ethics, then (2.2.2) might constitute a cogent way of defining morally-acceptable risks. As it is, however, in (2.2.2) Bethe and Rasmussen again beg the very question they attempt to answer. If society acts immorally in accepting risk *A*, then its decision to accept risk *B*, because it is of a lower magnitude than *A*, cannot be called moral. Yet, by (2.2.2), Bethe and Rasmussen appear to conclude that accepting *B* would be moral. On the other hand, if society acts morally in accepting risk *A*, then its decision to accept risk *B*, which is of lesser magnitude than *A*, cannot be called moral purely on grounds of the *magnitude* of the risk. Perhaps there are insufficient good reasons for taking risk *B*, perhaps the burden of risk *B* would not be borne equitably, or perhaps *A* was voluntarily chosen but *B* was involuntarily imposed.

2.2.2. *The Comparability of Voluntarily-Chosen and Involuntarily Imposed Risks*

Living near a nuclear plant is less a risk than driving an automobile, according to Bethe and Rasmussen. But is it just and equitable that *involuntary* risks (e.g., of nuclear accidents) be treated on the same level as *voluntary* ones (e.g., taken whenever one chooses to drive an automobile)? Is it not an open question whether voluntary and involuntary risks ought to be deemed equally acceptable? Is it not an open question whether society ought to force one to accept a risk comparable to that which one could reject as an individual? After all, technological decisions often involve a tradeoff between utility (usually economic) and risk (usually medical). Because of the tendency to maximize the importance of economic utility and to minimize that of individual well-being, should not one be much more careful in evaluating technologies (like nuclear power) not voluntarily chosen by every individual who bears the risk from them?[37] This insight, that the risk taken regarding voluntarily-chosen technologies is not comparable to that of involuntarily-imposed ones, is not discussed fully either by Bethe or by Rasmussen. Instead they appeal only to the magnitude of the risks they discuss, and fail to address the assumption that voluntarily-chosen hazards provide a criterion of what may be involuntarily imposed. For this reason, premise (2.2.2) is not, in general, true. It errs in defining an ethical notion in terms of a societal scientific one, viz., society's past choices. For this reason it commits the naturalistic fallacy.

In employing (2.2.2) Bethe and Rasmussen seem to have focused on mathematical and scientific conditions which are *necessary* for the successful ethical employment of nuclear energy, but have addressed, neither what might constitute correct *sufficient* conditions for this decision, nor why they are sufficient. They have only addressed the degree to which nuclear energy is economically beneficial and medically risky, rather than whether this risk ought to be taken. This, of course, is not really what is at issue in the continuing public policy debate over nuclear technology. As both proponents and opponents of nuclear energy maintain, "current debate over whether nuclear power is safe or unsafe emerges as a spurious issue, for both sides recognize that nuclear power is an inherently dangerous technology".[38] All those involved in the controversy "recognize the inherent danger of nuclear power plants".[39] If this is the case, then mathematical probabilities of the nuclear accident risk are not really the major issues in the ongoing debate regarding nuclear fission technology.

Even on mathematical grounds, however, perhaps Bethe and

Rasmussen can be faulted. The mathematical cornerstone of their argument in (2.2) is that the accident risk per year (the 'expected value' of the risk, given the probability of the accident and the number of fatalities for that event) is greater for fires, for example, than for nuclear reactors (Figure 2A, WASH-1400). When one realizes, however, that the expected risk value from reactors was calculated (Figure 2A) such that delayed cancer deaths and genetic deaths were excluded, data on the two risks become less meaningful. Since "delayed cancer fatalities exceed prompt" ones "by a factor of 100 or more", and since "the absolute number of genetic defects may be larger than the number of cancer deaths"[40] the nuclear risk data employed in Figure 2A of WASH-1400 are still questionable. One is yet able to ask a number of questions. Is any increased risk good? More importantly, is the nuclear risk good, once it is understood to include delayed as well as prompt genetic fatalities (in which case, contrary to the WASH-1400 interpretation, the risks of catastrophic reactor failure approach those of a variety of man-made and natural hazards)?

2.3. THE ARGUMENT BASED ON MAGNITUDE OF ACCI-
      DENT CONSEQUENCES

If Bethe's and Rasmussen's third major argument, (2.3), were translated into valid syllogistic form, it might be stated as follows:

(2.3.1) All major reactor accidents are technological and industrial catastrophes, each of which could result in a total of approximately 5000 additional cancer deaths.

(2.3.2) All technological and industrial catastrophes, each of which could result in a total of approximately 5000 additional cancer deaths, are accidents which could cause less than a 1.66% increase in the 300,000 cancer deaths per year.

(2.3.3) All accidents which could cause less than a 1.66%

increase in the 300,000 cancer deaths per year are insignificant risks which ought to be taken, in exchange for economic benefits.

(2.3.4) All major reactor accidents are insignificant risks which ought to be taken, in exchange for economic benefits.[41]

As in the two previous arguments, in this one Rasmussen and Bethe state explicitly only (2.3.1), (2.3.2), and (2.3.4). Since an argument of this form, i.e., *A* is *B, B* is *C*, therefore *A* is *D*, is logically invalid, it is reasonable to assume that Rasmussen and Bethe intended an enthymeme with suppressed premise (2.3.3). Taken together, (2.3.1), (2.3.2), (2.3.3), and (2.3.4) constitute a valid argument. Although (2.3.1) is doubtful because of the mathematical methodology involved,[42] let us assume, for purposes of argument, that (2.3.1) and (2.3.2) are true, since their verity has no bearing on whether the naturalistic fallacy is committed.

2.3.1. *The Moral Acceptability of 'Statistically Insignificant' Numbers of Induced Deaths*

The interesting premise, from the point of view of the naturalistic fallacy, is (2.3.3). Here moral acceptability is defined in terms of statistical significance, and an 'ought' proposition is deduced from an 'is' proposition. To understand why, in general, (2.3.3) is not true, consider the following example.

Since US burial of its radioactive wastes within one mile outside the established territorial waters of Great Britain would cause less than an 1.66% increase in cancer deaths per year, this risk ought to be taken.

Although substantial short-term economic benefits would accrue to the US from the burial, and the risks to anyone (especially the British) allegedly would not result in a great increase in cancer, premise (2.3.3) would lead us to believe, erroneously, that the behavior in question is morally desirable. Considerations of justice and equity, on the other hand, demand that the US bear responsibility for its wastes, that it not subject innocent people to any risk, however small, and that those who receive benefits from a particular technology ought also to be the ones who bear the burdens of it.

The problem with premise (2.3.3) is that if it is correct, then as *individuals*, human beings have no rights to life. Rather only statistically meaningful numbers of human lives appear to have a moral significance. But if only aggregates of lives have a moral significance, then the principle of individual rights is not being recognized.

2.3.2. *The Utilitarian Assumptions of the Argument*

The presence of a utilitarian framework is particularly obvious when one considers this assumption of (2.3.3), viz., that a statistically inconsequential increase in technology-induced human fatalities is morally insignificant. This assumption clearly is not based on considerations of protecting individual rights or safeguarding equity or equality among individuals; it is addressed, instead, to maximization of the overall good. In fact, only in the context of striving for the greatest good of the greatest number of people (apart from what individual rights might thereby be violated), is the death of a given number of people ever 'insignificant'. Hence (2.3.3) seems explicitly to point toward a utilitarian orientation of the Bethe-Rasmussen assessment.

As a number of authors have pointed out, to the extent that energy-policy decision-making is utilitarian, it shares the major weaknesses of this ethical theory. Some of these deficiencies include a general insensitivity to considerations of equity; a disregard for future generations; a tendency to equate desires with need; and an assessment only of quantifiable goods and bads. Because of its utilitarian orientation, it is also possible that the Bethe-Rasmussen analysis might encourage policymakers to sanction a system in which noneconomic benefits, e.g., aesthetic and health-related goods, received no consideration at all. A second possible consequence of acceptance of this utilitarian framework is that one might be led to seek a technology causing overall maximization of economic benefits even though economic inequities or hardships were visited upon a given minority. This latter consequence might occur as a result of the fact that a system of utilitarianism is, by definition, insensitive to demands based on equity and equal justice.[43]

Inasmuch as premise (2.3.3) provides a probabilistic criterion for assessing when a technological risk is morally significant, its problematic nature is shared by all other purely statistical criteria for moral acceptability. This difficulty is very likely a result of the fact that, in our society, law always comes *after* the potential for injury has been demonstrated. Probable injuries cannot be treated in courts of law. Since retroactivity is inherent in our legal system, it is not surprising that the *probability* of widespread injury or death, rather than the *actuality* of only a few such events, has been chosen as a criterion according to which a technology may be judged morally acceptable. Although this assumption may be reprehensible, there is at least one sense in which the moral blindness it perpetrates is part of a larger legal framework. Within this framework an action or a technology can be shown to be unethical only after positive demonstration of damages is made. Hence part of the problem is the historical genesis of a legal system based on the values and conflicts of the private sectors of society. As Michael Baram put it, "the courts have not been designed to serve as oracles or social planners, but to grapple with actual conflict manifested in specific acts or injuries".[44]

2.3.3. *What Is Normal as a Criterion for What Is Moral*

A final problem with premise (2.3.3) is that it is based on the assumption that the 'normal' number of deaths is moral or acceptable.[45] Because a normal number of cancers *is* occurring each year, Bethe and Rasmussen conclude that this number *ought* to occur.

In defining what is 'normal' as moral, e.g., a given cancer rate is normal and therefore moral or acceptable, Bethe and Rasmussen are holding a position leading to undesirable consequences. One such consequence is that the status quo will thereby be accepted as morally desirable. This consequence is particularly dangerous within technology assessment because it allows one to beg important questions and merely define them in terms of past answers. Two of the many questions thus begged by Rasmussen and Bethe are that the current rate of cancer deaths is normal and therefore acceptable, and that increasing energy consumption is normal and therefore acceptable. Although these assumptions are built into the status quo, failing to analyze them means that alternative answers to the energy dilemma are simply ignored. As a consequence, whatever social and ethical discrepancies that are present in the status quo are also present in the technology assessment.

In assuming that what is normal is moral, one is employing a philosophical assumption about what is good; this assumption goes uncriticized because it is masked as common-sensical, ordinary, or part of 'normal' life. Hence the views, of whatever people set the norm, enjoy an undeserved and therefore irrational power. To accept 'normal' standards in certain areas of technology assessment thus deludes one, says Alasdair MacIntyre, into thinking that one's activity presupposes no philosophical point of view. Not to recognize this, he claims, is "to make oneself the victim of bad philosophy".[46] In our 'normal' activities we often do not take account of the extent to which what is called 'good' or 'ethical' is a product of what is thought to be normal or necessary.[47]

3. The Consequences to Public Policy

Because it presupposes the correctness of utilitarianism and of the

assumption that what is normal is moral, the naturalistic definition
proposed in premise (2.3.3) is highly doubtful. Like the naturalistic
definitions employed in premises (2.1.2) and (2.2.2), this proposition is based on the assumption that certain scientific and mathematical facts constitute sufficient grounds for drawing an ethical
conclusion.

3.1. VIOLATIONS OF EQUITY AND ACCEPTANCE OF THE STATUS QUO

Apart from being methodologically doubtful or even erroneous,
commission of the naturalistic fallacy can lead to substantive,
undesirable consequences, both in the technology assessments
where it occurs and in the public policy based on these assessments. Several consequences of the Bethe-Rasmussen employment
of the fallacy include: (1) according a higher priority to meeting
energy demand and promoting economic growth than to preserving
basic rights, especially equal protection; (2) acceptance of the
status quo, or whatever is taken as 'normal', even if it may not be
morally desirable; and (3) allowing experts to make science-related
public policy judgments rather than also providing for the public
to have a voice in these determinations and, therefore, permitting
these judgments to deal only with scientific parameters rather than
also with ethical ones.

The first consequence was discussed in connection with evaluation of argument (2.1) in Section 2.1.[48] It is clear from that
analysis that if the magnitude of risk (or any other purely empirical
criterion) involved in a particular, economically-beneficial technology is a sufficient basis for determining its moral acceptability,
then widespread violations of rights or principles of equity are
likely to occur when that technology is implemented. The Bethe-
Rasmussen analysis fails to preclude such consequences precisely
because it avoids even the mention of values such as freedom,

justice, and equity, which might be jeopardized as a result of policy regarding a particular technology.

The second consequence was discussed in Sections 2.1 and 2.3.[49] From this analysis it is clear that so long as the status quo is taken as the norm for what is morally acceptable, then public policy making will beg the very questions it is supposed to answer. It also will be based on the false assumption that consistency is a sufficient condition for morally-acceptable judgments.

3.2. DENIAL OF CITIZENS' ROLES IN POLICY-MAKING

With respect to public policy, the third consequence (that technology-related decision-making will fail to include considerations, both of nonscientific parameters and of the rights of the public to a voice in these matters), is one of the most important results of the Bethe-Rasmussen commission of the naturalistic fallacy. In reducing ethical concerns to purely mathematical and scientific (including economic) terms, Bethe and Rasmussen have argued as if what were at issue were acceptance of purely technical considerations. If this were the case, the layman would have no right to decide questions involving such subject matters; therefore as Bethe and Rasmussen have defined the issue of nuclear technology, it is one that must be left to an elite group of intellectuals, to 'experts', not to the public, to decide.

Interestingly the widespread acceptance of the naturalistic fallacy, as employed by Rasmussen, Bethe, and others, has almost certainly already led to the suggestion that the public ought to leave the nuclear energy decision to scientists and mathematicians. In fact one of the top officials, of the governmental agency regulating nuclear power, said recently that he wants "to eliminate where possible from the public debate over nuclear energy extraneous arguments which cloud" the issue. The extraneous concerns which he called "irrelevant" and "phony" include such questions as whether government regulations provide enough voice for consumer, as well as utility, interests; whether the government has stifled opposition to atomic energy; and whether it has disclosed all scientific documents relevant to the safety of nuclear power plants.[50] All the issues mentioned by him as irrelevant involve questions of ethics, equity, or justice. What his attitude and consequence (3) both suggest is that the public might have a difficult time getting a hearing in the nuclear debate. This consequence is particularly dangerous to public policy for a number of reasons. Acceptance of (3) means that it is possible that a new class of elitists might be created to control technology and its consequences. By "carefully constructing a climate where the only decisions that appear rational to decision makers are those that are already tacitly decided" on purely scientific or economic grounds, matters important to the public interest might not be considered.[51]

This means not only that key ethical, legal, social, political, and psychological aspects of policy-making regarding technology might be ignored, but also that these decisions might not be arrived at in a democratic manner. As Juvenal put it: *"Quis custodiet ipsos custodes?"* "Who will watch the watchers themselves?" Only society as a whole is able to watch the custodians of policy regarding technology. Yet paradoxically this is precisely what would not be possible if one accepts the presuppositions of the Bethe-Rasmussen commission of the naturalistic fallacy. This is probably why Macpherson warned:

"technological change in our lifetime, if left to operate by itself within the present social structure and guided only by our present ambivalent ontology, ... is as likely to prevent as to promote the realization of liberal democratic ends."[52]

If Macpherson's analysis is correct, then contrary to assumptions built into naturalistic ethics, public policy regarding technology ought not be made by scientists alone. Although technical experts

must rightfully discuss *how* a particular technology can be implemented, no person is better able than another to judge *whether*
it ought to be implemented. As Muchnicki put it, the key issues
regarding regulation of technology involve "social policy" and not
"technical expertise"; the "central question . . . is what society
really wants".[53] In other words, deciding to accept a technological
risk "is an entirely human judgment, that has nothing to do with
whether you're a farmer or an engineer or a mathematician".[54]

Another reason why certain technology-related decisions ought
to be made, in part, by the public rather than solely by technical
experts is that they involve political interests and pressure-group
power. Sometimes the very scientists and engineers who know the
most about a particular technology are those who are employed
by one political interest group prominent in the technology debate.
Although such conditions of employment do not mean that the
experts in question are dishonest or biased, they do suggest that
these persons may have become accustomed to a particular way of
thinking which ignores important considerations. This is why Price
maintains that "the freedom of scientific institutions depends
on" the recognition that "the scientific mode of knowledge is
inherently limited in its ability to deal with major questions of
public policy".[55] In the case of nuclear fission technology, for
example, there allegedly have been instances in which scientists
and engineers refused to testify regarding nuclear safety because
they feared cutoff of research support from a major utility or
from pro-nuclear government funding agencies.[56] Perhaps one of
the most obvious signs of this apparent conflict of interest is that
so many scientists and engineers have found it necessary to resign
from their posts in order to assess nuclear technology in a way
they believe is objective.[57]

Looking at this problem from the vantage point of the nuclear
plant licensing process, one attorney made an interesting remark.
He said that in the hearings, nearly all experts are in the pay of the

AEC or the utilities, leaving only "mavericks, students, and a few conscience-ridden academics" to speak on the side of the public.[58] Moreover the current government agency (the Nuclear Regulatory Commission) has admitted tampering with 'scientific' data regarding ill effects of nuclear power plants.[59] Even the famous Rasmussen Report, the only allegedly complete study of nuclear power safety, was directed and staffed by persons who, up to that time, had been under contract to a government agency (the AEC) whose purpose was "to promote" nuclear energy.[60] While such conditions, alone, in no way invalidate the results of this report, it is obvious that the most objective assessment of nuclear technology would probably be done by an agency whose purpose was neither to promote nor to condemn nuclear power.

It should not be surprising that the public policy issues surrounding nuclear technology are political as well as scientific. After all, in addition to utility monies, "more than a hundred billion dollars [in federal funds] has already been invested in nuclear power" subsidies since 1945.[61] Hence it is only logical that both government and industry should wish to guarantee a return on their investment and to provide inexpensive and abundant energy for all consumers. Especially for this reason, however, decisions regarding nuclear technology ought not to be made by scientists alone. But if this is the case, then employment of the naturalistic fallacy ought to be avoided in technolgy assessments.

4. New Directions for Technology and Public Policy

If my remarks about the Bethe-Rasmussen analyses are correct, then future evaluations of technology ought to be improved in significant ways. Specifically they ought to focus explicitly on: (1) the moral acceptability, rather than merely on the magnitude, of technological benefits and risks; (2) the ethical constraints operative in deciding to implement an involuntarily-imposed (as opposed to a voluntarily-chosen) technology; (3) the moral significance of "statistically insignificant" increases in technology-induced fatalities; and (4) the value presuppositions hidden in employing what is 'normal' or 'necessary' as a criterion for acceptable policy regarding technology.

As a result of my analysis, it also appears that technology studies used to undergird public policy need to give adequate attention to a number of problems. (1) Their methodology and scope is often not wholistic, but narrowly technical.[62] (2) Considerations of individual rights and equal protection under the law frequently have been displaced by economic and utilitarian concerns. (3) Democratic control of public policy-making often has been stifled by lack of public participation, by failure to employ an 'adversary system' of evaluation of technology, and by forgetting the political dimensions of allegedly scientific or technical issues.

What these problems suggest is not that proponents of a particular technology are always unprincipled persons attending only to considerations of economics and expediency. On the contrary, I believe that most supporters of nuclear energy, for example, are honest persons attempting to do what is right. Because of the complexity of most science-related political issues, because of the unduly emotive stance of many opponents of technology, and because of the narrowly-technical training of many persons responsible for science-related public policy, however, important ethical and social parameters often are omitted from consideration. The problem, I think, is that decision makers are directly or indirectly pressured by obvious factors such as energy demand and, as a result, lose sight of the more subtle or long-term consequences of their policy choices. Moreover when a new technology is implemented, its first proponents do not have the same benefit of hindsight, so often employed years later by its opponents who are able to discover grounds for criticizing its use.

Even if hindsight is not always available when one is attempting to evaluate science-related public policy, implementing the four suggestions for reform and avoiding the three problem areas I have outlined are a first step toward foresight. Meanwhile, as Chapter 5 suggests, we can begin to compensate the public for the social costs which are inequitably and involuntarily imposed on them by nuclear fission and by other technologies. These suggestions, however, describe only part of the task facing makers of public policy. The real need, a much larger one, is to aim at the political ideal of the philosopher-king. If philosophers were 'kings', or if philosophical considerations were included in democratic decision-making, then public policy might be both socially and scientifically sound. To paraphrase Plato:

> Until philosophers are the makers of public policy, or the public policy makers of this world have the spirit and power of science, and political greatness and wisdom met in one, neither states nor the human race itself will ever have rest from their evils.[63]

Notes

[1] This same point is made by A. MacIntyre, 'Utilitarianism and Cost-Benefit Analysis', in *Values in the Electric Power Industry* (ed. by Kenneth Sayre), University Press, Notre Dame, 1977, p. 217; hereafter cited as: *Values.*
[2] Moore, *Principla Ethica*, University Press, Cambridge, 1951, pp. viii-ix; hereafter cited as: *PE.*
[3] *PE*, p. 60.
[4] *PE*, p. 39.
[5] *PE*, p. 60; see also W.K. Frankena, 'The Naturalistic Fallacy', *Mind* 48 (192), (October 1939), 467; and Abraham Edel, 'The Logical Structure of Moore's Ethical Theory', in *The Philosophy of G.E. Moore* (ed. by P.A. Schilpp), Northwestern University Press, Chicago, 1942, p. 146; hereafter cited as: *Moore.*
[6] *PE*, p. 39. In a more general sense, the naturalistic fallacy may be said to be a variant of the definist fallacy. This latter error is committed whenever one attempts "to reduce one simple notion to another". (J.H. Olthuis, *Facts, Values, and Ethics*, Van Gorcum, Assen, 1969, p. 2; hereafter cited as: *Facts.* See Frankena, 'The Naturalistic Fallacy', pp. 464–77; and George Nakhnikian, 'On the Naturalistic Fallacy', in *Studies in the Philosophy of G.E. Moore* (ed. by E.D. Klemke), Quadrangle, Chicago, 1969, p. 65.) In its specific sense, however, the naturalistic fallacy is the attempt to define values

in terms of psychological or empirical facts. Although Moore initially (1903) confused the generic and specific variants of the fallacy, he admitted in an unpublished draft (intended as the Preface to the second edition of *Principia Ethica* in 1922) that the specific sense was what he had intended all along. He later came to believe that this is the only respect in which one could be said to have committed the Naturalistic Fallacy. (Olthuis, *Facts*, pp. 30, 34.) For this reason my references to the error are to the specific rather than the generic variant of it.

[7] Moore, *PE*, p. 40, says that the fallacy "consists in substituting for 'good' some one property of a natural object or of a collection of natural objects; and in thus replacing ethics by some one of the natural sciences".

[8] Searle says that the naturalistic fallacy is committed whenever one attempts to derive an 'ought' from an 'is'. See J.R. Searle, 'How To Derive "Ought" from "Is"', *Philosophical Review* **73** (1), (January 1964), 43–58; hereafter cited as: 'Ought'.

[9] Following Kohlberg, Giarelli maintains: "To fail to consider the open question is to commit the naturalistic fallacy". See J.M. Giarelli, 'Lawrence Kohlberg and G.E. Moore', *Educational Theory* **26** (4), (Fall 1976), 350.

[10] *PE*, pp. 23–24. See also p. 36. It should be noted that although Moore argues that ethical judgments ought not be reduced to purely scientific ones, he does not deny that causal or empirical propositions are a part of ethics. Moore is definitely not a value skeptic. See F. Snare, 'Three Sceptical Theses in Ethics', *American Philosophical Quarterly* **14** (2), (April 1977), 129–130; hereafter cited as: 'Theses'.

[11] *PE*, p. 108.

[12] *PE*, p. 73.

[13] Although variant (2) of the naturalistic fallacy seems similar to (1) in that the essential problem of both is the attempt to reduce ethical to non-ethical characteristics, the is/ought error has been thought by some to be different from the naturalistic fallacy. In particular, a number of philosophers have questioned whether the fallacy, as described by Moore, includes the attempt to derive 'ought' from 'is'. Bruening, Frankena, Snare, and White see the two problems as not essentially related, while Kohlberg, Searle, Stevenson, Waddington, Murdoch, Nowell-Smith, Veatch, Prior, Giarelli, Hare, and Flew all believe that the naturalistic fallacy is committed whenever one attempts to derive an 'ought' from an 'is'. (See W. H. Bruening, 'Moore and "Is-Ought"', *Ethics* **81** (2), (January 1971), 143–49, esp. pp. 146-47; Giarelli, 'Lawrence Kohlberg', p. 353; Kohlberg and R. Mayer, 'Development as the Aim of Education', *Harvard Educational Review* **42** (4), (1972), 466; L. Kohlberg, 'From Is to Ought: How To Commit the Naturalistic Fallacy and Get Away with It in the Study of Moral Development', in *Cognitive Development and Epistemology* (ed. by T. Mischel), Academic Press, New York, 1971, p. 154; Olthuis, *Facts*, pp. 28–34; and Searle, 'Ought', pp. 43–58.

Interesting as this question of the is/ought issue is, no further analysis of its relation to the naturalistic fallacy will be made here, since the main purpose of this chapter is to evaluate the ethical assumptions built into some recent technology assessments of atomic energy. However, apart from whether the is/ought issue is part of what Moore meant by the naturalistic fallacy, it is clear that others have described the error in this manner, and that the attempt to derive an 'ought' from an 'is' constitutes a substantial problem in ethics. Even Frankena, who criticized Moore's notion of the naturalistic fallacy, agrees that the is/ought error should be avoided. He maintains that "what makes ethical judgments seem irreducible to natural or to metaphysical judgments is their apparently normative character". (W.K. Frankena, 'Obligations and Value in the Ethics of G.E. Moore', in *Perspectives on Morality*: *Essays by William K. Frankena* (ed. by K.E. Goodpaster), University Press, Notre Dame, 1976, p. 16; hereafter cited as: 'Perspectives'.

14   See note 9.

15   E. H. Duncan, 'Has Anyone Committed the Naturalistic Fallacy?', *Southern Journal of Philosophy* **8** (1), (Spring 1970), 49–50; hereafter cited as: 'Fallacy'.

16   The whole point of the open question is to challenge the analyticity of definitions of good. If they are not analytic, then it is fallacious to hold that propositions involving ethical terms are analytic, tautologous, or true by definition. Consequently it is also unwise, if not erroneous, for one to fail to consider an argument which might challenge his particular definition of good and for one to treat this definition as 'closed'. For further information regarding the open question, see *PE*, pp. 15, 38; Roger Hancock, 'The Refutation of Naturalism in Moore and Hare', in *Studies* (ed. by E.D. Klemke), p. 45; Snare, 'Theses', p. 136.

17   W.K. Frankena, 'Concluding More or Less Philosophical Postscript', in *Perspectives on Morality* (ed. by K.E. Goodpaster), University Press, Notre Dame, 1976, p. 210. See also Duncan, 'Fallacy', p. 49; and Olthuis, *Facts*, pp. 25–27.

18   Olthuis, *Facts*, p. 26.

19   Frankena, 'The Naturalistic Fallacy', pp. 468–75, claims that the error ought to be avoided because it is a form of moral blindness in which one does not see the unique *normative* character of ethical statements. See Snare, Theses, pp. 129–36, however, for an argument that commission of the naturalistic fallacy is a logical error. Despite different opinions as to the nature of this fallacy, most philosophers maintain that it ought to be avoided. (See Frankena, 'Concluding More or Less Philosophical Postscript', in *Perspectives* (ed. by K.E. Goodpaster), p. 210; E.F. Walter, 'A Defense of Naturalism', *The Journal of Value Inquiry* **7** (3), (Fall 1973), 220; Duncan, 'Fallacy', p. 49; and Olthuis, *Facts*, p. 26.

[20] *PE*, p. 20. See also H.J. Paton, 'The Alleged Independence of Goodness', in *Moore* (ed. by Schilpp), p. 133; and Olthuis, *Facts*, p. 29.

[21] *PE*, pp. 20 and 40. As G.E. Moore puts it, if we do not accept commission of the naturalistic fallacy, "we shall be much more apt to look about us". (*PE*, p. 39.)

[22] US Nuclear Regulatory Commission, *Reactor Safety Study – An Assessment of Accident Risks in US Commercial Nuclear Power Plants*, Report No. WASH-1400, NUREG-75/014, Government Printing Office, Washington, D.C., 1975; hereafter cited as: 'WASH-1400'. H.A. Bethe, 'The Necessity of Fission Power', *Scientific American* **234** (1), (January 1976), 21–31; hereafter cited as: 'Fission'.

[23] This question is the topic of 'WASH-1400' and is discussed on pp. 25–27 of Bethe, 'Fission'.

[24] Bethe, 'Fission', p. 25; US Nuclear Regulatory Commission, 'WASH-1400', pp. 157, 195. See also Summary Report of the US Atomic Energy Commission, 'Reactor Safety Study: An Assessment of Accident Risks in US Commercial Nuclear Power Plants', *Atomic Energy Law Journal* 16 (3), (Fall 1974), 177–204; hereafter cited as: AEC, 'Summary Report'. Accident probabilities are given in Section 2.1.1. According to the report (p. 178), "the risks to the public from potential accidents in nuclear power plants are very small".

Although the point of this essay is not to evaluate the truth of the premisses in any of these arguments, there is reason to doubt two of Bethe's claims (p. 25) made in connection with premise (2.1). He says that there has never been a core melt and that, in the event of a core melt, radioactivity would not be released. Regarding the former claim, it is important to note that there was an explosion and partial core melt in 1961 at the Idaho Falls (light-water) reactor, where three persons were killed. There was also a partial core melt of the Fermi (fast breeder) reactor in 1966 in Detroit. Both of these accidents caused releases of radiation into the atmosphere. (These and other accidents are documented in J.J. Berger, *Nuclear Power*, Dell, New York, 1977, pp. 40–43; James Elder, 'Nuclear Torts: The Price-Anderson Act and the Potential for Uncompensated Injury', *New England Law Review* **11** (1), (Fall, 1975), 130ff.; J.G. Fuller, *We Almost Lost Detroit*, Ballantine, New York, 1975, pp. 17, 86–90, 109–25, 256–57; James Lieberman, 'Generic Hearings: Preparation for the Future', *Atomic Energy Law Journal* **16** (2), (Summer 1974), 161ff.; J. Marrone, 'The Price-Anderson Act: The Insurance Industry's View', *Forum* **12** (2), (Winter 1977), 610ff.; S. Novick, *The Electric War*, Sierra, San Francisco, 1976, pp. 201–281; J. Primack and F. Von Hippel, 'Nuclear Reactor Safety', *The Bulletin of the Atomic Scientists* **30** (8), (October 1974), 5–12; and M.A. Rowden, 'Nuclear Power Regulation in the United States', *Atomic Energy Law Journal* **17** (2),

(Summer 1975), 106–109.) Fuller also cites Bethe's claim, that the Fermi accident was "incredible and impossible" (*Detroit*, p. 221).

Bethe's second claim, that radioactivity would not be released in the event of a core melt, has also been disputed. Critics point to the accidents mentioned above, and to the fact that the emergency core cooling system (which should prevent such a release of radiation) has never been tested. At least three facts are significant in this regard. First, no full-scale empirical tests of the ECCS have ever taken place (Novick, *Electric War*, pp. 155–56; Primack and Von Hippel, 'Nuclear', pp. 7–9; M. Bauser, 'United States Nuclear Export Policy', *Harvard Internatonal Law Journal* 18(2), (Spring 1977), 51.) Secondly, all of the small, scale-model tests (six of six) of the ECCS have failed. (Primack and Von Hippel, 'Nuclear', pp. 7 and 9. J.J. Berger, *Nuclear Power*, pp. 50–52.) Thirdly, the full-scale-model tests would be "possibly very destructive" and nuclear proponents are unwilling to take this risk (Nuclear proponents F.H. Schmidt and D. Bodansky, in *The Energy Controversy*, Albion, San Francisco, 1976, pp. 139–42, explain clearly that high risk is the reason for failure to conduct proper ECCS tests. Novick, *Electric War*, p. 192, also quotes industry and government leaders who subscribe to the same explanation as that given by Schmidt and Bodansky.)

[25] Bethe, 'Fission', p. 27 and US Nuclear Regulatory Commission, 'WASH-1400', pp. 28–29, 187–223; AEC, 'Summary Report', esp. p. 179. In Sections 1 and 2.14 of the report, the Rasmussen Committee makes this argument. They compare risks from nuclear accidents to those from other human-caused (e.g., automobile accidents) and natural events (e.g., tornadoes). See Figures 1, 2, and 3, and Table 1 in Section 1, 'WASH-1400', for further information regarding argument (2.2). Bethe's claim, that the risk from nuclear reactors is very small, will not be discussed here, since the point of this discussion is the ethical methodology he employs, rather than the factual correctness of his claims. For an alternative view of the cancer and genetic damage resulting from a nuclear plant accident, see C. Hohenemser, R. Kasperson, and R. Kates, 'The Distrust of Nuclear Power', *Science* **196** (4285), (April 1977), 25–34. Hohenemser *et al.* explain what methodological errors in the Bethe position are responsible for the under-estimation of the cancer and genetic damage risk to the public from reactors. Significantly higher risk estimates are also found in a government document known as 'The Brookhaven Report', i.e., 'Theoretical Possibilities and Consequences of Major Accidents in Large Nuclear Power Plants', USAEC Report WASH-740, 1957. According to this report, updated in 1965 (R.J. Mulvihill, D.R. Arnold, C.E. Bloomquist, and B. Epstein, 'Analysis of United States Power Reactor Accident Probability', PRC R-695, Planning Research Corporation, Los Angeles, 1965; this unpublished draft is from the file 'WASH-740 update', Public Documents Room, Nuclear Regulatory Commission.), the Bethe-

Rasmussen statistics would need to be revised upward, in various instances, by factors of 30 to 100.

[26] AEC, 'Summary Report', pp. 182–83; see also Section 2.16 of the report NRC, 'WASH-1400', p. 247. As Bethe (p. 26) puts it: "A reactor accident clearly would not be the end of the world, as many opponents of nuclear power try to picture it. It is less serious than most minor wars."

Just as the Bethe-Rasmussen statistics cited in (2.2) are contrary to those found in WASH-740, so also those employed in argument (2.3) are quite different from those in WASH-740. Whereas Bethe and Rasmussen maintain that a total of 5000 cancer *deaths* would result both immediately and in the years after a nuclear accident, the Brookhaven Report estimates immediate deaths at 45,000 and early illnesses (with later deaths) at 100,000. (See note 25 above.) A summary of these (Brookhaven) statistics is given in James Elder, 'Nuclear Torts', pp. 111–135, esp. p. 127. See also Hohenemser *et al.*, 'Distrust', pp. 28–34, for a mathematical and methodological analysis of these risk estimates. Hohenemser explains that some of the Rasmussen-Bethe fatality estimates are lower than the Brookhaven-APS figures because Rasmussen and Bethe exclude all deaths caused by a nuclear accident, but which are delayed (e.g., cancers), and all genetic damages likewise caused by an accident, but which appear only later (e.g., in many successive generations). According to Hohenemser *et al.*, Bethe and Rasmussen are wrong in denying that the "risks of catastrophic reactor failure approach the risks of a variety of man-made and natural catastrophic hazards" (p. 29). The risks are about the same, they argue, because delayed nuclear accident fatalities exceed prompt ones by a factor of 100 or more; when the delayed deaths are considered, the nuclear risk rises sharply.

[27] See note 24. According to the US Environmental Protection Agency, the 'risk approach' presented in WASH-1400 implies "an acceptability judgment [regarding nuclear power] to the average reader". (NRC, 'WASH-1400', Appendix XI, p. 2-2.)

[28] See Frankena, 'The Naturalistic Fallacy', p. 468, who uses a similar approach to an argument employed by Mill.

[29] Both the Brookhaven (WASH-740) statistics, as well as a more complete mathematical analysis, plus past inductive experience, suggest that this number is inaccurate. See notes 24, 25, and 26 for this information, including the statistics given in the Brookhaven Report, WASH-740. The Rasmussen probability estimates are also misleading, unless they are placed in a more complete mathematical context. Both Bethe and Rasmussen point out that a core melt accident has a probability of about one in 17,000 per reactor per year. While correct, this statistic gives only a year-by-year probability of a core melt for *one* reactor. Using the same Rasmussen data, it is more meaningful to compute the probability that a core melt will occur in one of the 75

plants now operating during their 30-year lifetime. As was pointed out in Chapter Four, Section 2.2.1. this probability is 12%. Likewise the probability of a core melt in one of the reactors now operating or under construction is 25%. Given this more accurate mathematical data regarding core melt probability, it is somewhat misleading for Bethe and Rasmussen to provide only single-year, single-reactor probabilities, and then to conclude in (2.1.1) that nuclear accidents have a "very low probability".

Related evaluations of the design adequacy of WASH-1400, including discussion of fault-tree analysis, completeness of the parameters considered, and mathematical extrapolation in the study, may be found in Hohenemser *et al.*, 'Distrust', pp. 28–34. See note 24 for information regarding previous nuclear accidents.

30 The APS results are consistent with those of the Brookhaven Report. Moreover the APS study was done at the same time as the Rasmussen Report; it was performed by a group of the most prestigious nuclear scientists in the world, while the Rasmussen Report was done by nuclear proponents under AEC contract. The APS explained why WASH-1400 was able to reduce predicted catastrophic effects of a nuclear accident, viz., it ignored relevant data treated in WASH-740. Because of the mathematical errors inherent in the obsolete 'fault-tree analysis' of the Rasmussen Report, and because of a number of implausible assumptions (e.g., 90% evacuation; downward fuel melt; no incidence of sabotage; no irradiation of special tissues; downwind radiation lasting only one day; no resource contamination through land and water, etc.), the APS rejected the Rasmussen Report. In all, the independent American Physical Society estimates of deaths, cancers and genetic damages arising from a nuclear accident are as much as 100 times greater than those given in the Rasmussen Report. See H.W. Lewis *et al.*, 'Report to the American Physical Society by the Study Group on Light-Water Reactor Safety', *Reviews of Modern Physics* **47** (1), (Summer 1975), S1–S124; and Study Group on Light-Water Reactor Safety, 'Nuclear Reactor Safety – the APS Submits Its Report', *Physics Today* **128** (7), (July 1975), 38–43. See also Hohenemser *et al.*, 'Distrust', pp. 28–34, as well as notes 24, 25, and 26 above. Hohenemser *et al.*, point to some of the same factors as do the authors of the APS report. They conclude that the risk of a core melt is calcuated as low, only because of the incompleteness of the mathematical parameters considered, and because of the omission of possible accident causes established through fault-tree analysis.

31 Since the Price-Anderson Act limits the liability of the nuclear industry to $ 560 million in damages for any one accident, although government estimates of losses for a single nuclear incident go as high as $ 17 billion (according to the Brookhaven Report, WASH-740), this means that as much as 97% of the damages resulting from a nuclear accident might not

be covered. For more information on this legislation, see Chapter Four.

[32] Given the limit on liability in the event of a nuclear accident, persons living near a power plant are those most likely to sustain uninsurable losses to their persons and property. This means that a disproportionate share of the costs of nuclear power are borne by those within a particular vicinity, and that discrimination on the basis of geographical considerations might occur. The Fifth and Fourteenth Amendments, on the other hand, apply equally to all persons in the US and prohibit any form of discrimination regarding "life, liberty, or property". Some persons have argued that the Price-Anderson liability limit represents a violation of property rights guaranteed through these two amendments. See Chapter Four, Section 2.3.2, for discussion of this issue.

[33] Bethe, 'Fission', p. 31.

[34] *PE*, p. 58; see also p. 43.

[35] For comparative figures regarding per capita energy consumption, see D. Meadows *et al.*, *The Limits to Growth*, New American Library, New York, 1974, pp. 63–78. See also W.T. Blackstone, 'Equality and Human Rights', *Monist* 52(4), (October 1968), 616–39; D.G. Kozlovsky, *An Ecological and Evolutionary Ethic*, Prentice-Hall, Englewood Cliffs, New Jersey, 1974, pp. 75–76, 101–105; G. Tyler Miller, *Living in the Environment*, Wadsworth, Belmont, 1975, pp. 7–21, 201–45; and A.M. Weinberg and R.P. Hammond, 'Global Effects of Increased Use of Energy', *Bulletin of the Atomic Scientists* 28 (3), (March 1972) 5–8. See also Ezra Mishan, *Technology and Growth: The Price We Pay*, Praeger, New York, 1969; and E.F. Schumacher, *Small Is Beautiful*, Harper and Row, New York, 1973.

[36] See notes 25 and 27.

[37] C. Starr, 'Social Benefit Versus Technological Risk', in *Technology and Society* (ed. by Noel de Nevers), Addison-Wesley, London, 1972, pp. 214–17, makes this same point.

[38] J.G. Palfrey, 'Energy and the Environment', *Columbia Law Review* 74 (8), (December 1974), 1377; hereafter cited as: 'Energy'.

[39] James Lieberman, 'Generic Hearings', *Atomic Energy Law Journal* 16 (2), (Summer 1974), 142.

[40] Hohenemser *et al.*, 'Distrust', pp. 28–29.

[41] What Bethe and Rasmussen also ask us to do, in this argument, is to look at the nuclear risk "in perspective" of other causes of death. As Bethe puts it, "One should remember that in the US there are more than 300,000 deaths every year from cancers due to other causes. A reactor accident clearly would not be the end of the world" since it would involve only approximately 5000 cancer deaths. "It is less serious than most minor wars." (Bethe, 'Fission', p. 26. AEC, 'Summary Report', pp. 182–83.) Further details regarding this argument (2.3) may be found in notes 26 and 27.

[42] See notes 24–30 as well as the preceding section of this chapter.

[43] For further discussion of utilitarianism, see Chapter Two, Sections 3.1 and 3.2. See also Alasdair MacIntyre, 'Utilitarianism and Cost-Benefit Analysis: An Essay on the Relevance of Moral Philosophy to Bureaucratic Theory', in *Values in the Electric Power Industry* (ed. by K.M. Sayre), University Press, Notre Dame, 1977, pp. 217–37; Sayre and Goodpaster, 'An Ethical Analysis of Power Company Decision-Making', in *Values* (ed. by Sayre), pp. 238–88.

[44] Baram, 'Social Control of Science and Technology', in *Philosophical Problems of Science and Technology* (ed. by A.M. Michalos), Allyn and Bacon, Boston, 1974, p. 527; hereafter cited as: 'Control'.

[45] See note 34.

[46] 'Utilitarianism', p. 217.

[47] See, for example, D.C. Anderson, 'Policy Riddle: Ecology vs the Economy', in *Environment and Society* (ed. by R.T. Roelofs, J.N. Crowley, and D.L. Hardesty), Prentice-Hall, Englewood Cliffs, 1974, p. 148.

[48] See notes 31 and 32.

[49] See note 34.

[50] W.O. Doub, 'Meeting the Challenge to Nuclear Energy Head-On', *Atomic Energy Law Journal* 15(4), (Winter 1974), 261, 263.

[51] D.J. Rose, 'New Laboratories for Old', in *Science and Its Public: The Changing Relationship* (ed. by G. Holton and W. Blanpied), D. Reidel, Dordrecht, 1976, p. 151; hereafter cited as: 'Laboratories'.

[52] C.B. Macpherson, 'Democratic Theory: Ontology and Technology', in *Philosophy and Technology* (ed. by Carl Mitcham and Robert MacKey), Free Press, New York, 1972, p. 161.

[53] E.D. Muchnicki, 'The Proper Role of the Public in Nuclear Power Plant Licensing Decisions', *Atomic Energy Law Journal* 15(1), (Spring 1973), 55, 59; hereafter cited as: 'Public'. Another way of making this same distinction is to point out that although technicians alone are able to determine the *degree* of risk and benefit inherent in a particular technology, only the public is able to evaluate whether it wants to *accept* this risk.

[54] Novick, *Electric War*, pp. 318–19.

[55] D.K. Price, 'Money and Influence: The Links of Science to Public Policy', in *Science* (ed. by Holton and Blanpied), p. 111.

[56] Novick, *Electric War*, p. 109; Baram, 'Control', pp. 531–32.

[57] Novick, *Electric War*, p. 256; Primack and Von Hippel, 'Nuclear', p. 9; G.C. Coggins, 'The Environmentalist's View of AEC's "Judicial" Function', *Atomic Energy Law Journal* 15(3), (Fall 1973), 184; and M.S. Young, 'A Survey of the Governmental Regulation of Nuclear Power Generation', *Marquette Law Review* 59(4), (1976), 847–48.

[58] Coggins, 'View', p. 187.

[59] V. McKim, 'Social and Environmental Values in Power Plant Licensing', in *Values* (ed. by Sayre), p. 54. See also Palfrey, 'Energy', pp. 1392, 1398–99; Coggins, 'View', p. 189; J. Elder, 'Nuclear Torts', p. 131.

[60] See AEC, 'Summary Report', p. 184.

[61] Berger, *Nuclear Power*, p. 112; see also pp. 94–97, 106, 144–47; and Primack and Von Hippel, 'Nuclear', p. 7; Muchnicki, 'Public', p. 45. Most of this money has come from the federal government, in terms of subsidies for research and development, waste disposal, fuel reprocessing, the Price-Anderson Act, and enrichment plants. Nevertheless some of it has also been provided by the utility industry.

[62] Rose, 'Laboratories', p. 152, points out in this regard that "respectable, clear-cut disciplines like nuclear engineering flourish at the expense of the large and difficult sociotechnical problem of how best to respond to the country's energy demands". For this reason, technology studies are often erroneously supposed to include evaluations only of scientific feasibility, technological and industrial feasibility, and economic and commercial feasibility. This is, in fact, the scope of the Bethe-Rasmussen analysis. Instead as Hafele points out, the notion of what must be included in the technology study should also include social-psychological-ethical feasibility. (See P.D. Pahner, 'Psychological Perspective of the Nuclear Energy Controversy', International Institute for Applied Systems Analysis, Vienna, Research Memorandum RM-76-67, 1967, p. 18.)

[63] Paraphrase of Plato, in the Seventh Letter to Dion and his family at Syracuse.

Name Index

Subject Index

PALLAS PAPERBACKS

Pallas Paperbacks Series is a natural outgrowth of Reidel's scholarly publishing activities in the humanities, social sciences, and hard sciences. It is designed to accommodate original works in specialized fields which, by nature of their broader applicability, deserve a larger audience and lower price than the standard academic hardback. Also to be included are books which have become modern classics in their fields, but have not yet benefitted from appearing in a more accessible edition.

Volumes appearing in Pallas will be promoted collectively and individually to appropriate markets. Since quality and low price are the two major objectives of this program, it is expected that the series will soon establish itself in campus bookstores and other suitable outlets.

PALLAS titles in print:

1. Wolff, *Surrender and Catch*
2. Fraser (ed.), *Thermodynamics in Geology*
3. Goodman, *The Structure of Appearance*
4. Schlesinger, *Religion and Scientific Method*
5. Aune, *Reason and Action*
6. Rosenberg, *Linguistic Representation*
7. Ruse, *Sociobiology: Sense or Nonsense?*
8. Loux, *Substance and Attribute*
9. Ihde, *Technics and Praxis*
10. Simon, *Models of Discovery*
11. Murphy, *Retribution, Justice, and Therapy*
12. Flato *et al.* (eds.), *Selected Papers (1937–1976) of Julian Schwinger*
13. Grandy, *Advanced Logic for Applications*
14. Sneed, *The Logical Structure of Mathematical Physics*
15. Shrader-Frechette, *Nuclear Power and Public Policy*
16. Shelp (ed.), *Justice and Health Care*
17. Petry, *G. W. F. Hegel. The Berlin Phenomenology*
18. Ruse, *Is Science Sexist?*
19. Castañeda, *Thinking and Doing*
20. Hilpinen (ed.), *Deontic Logic: Introductory and Systematic Readings*
21. Lehrer and Wagner, *Rational Consensus in Science and Society*
22. Bunge, *Scientific Materialism*